JOURNAL

OF

A RESIDENCE AND TOUR

IN THE

REPUBLIC OF MEXICO

JOURNAL

OF

A RESIDENCE AND TOUR

IN THE

REPUBLIC OF MEXICO

IN THE YEAR 1826.

WITH SOME ACCOUNT OF THE

MINES OF THAT COUNTRY.

By Capt. G. F. LYON, R.N. F.R.S.

IN TWO VOLUMES.
VOL. I.

KENNIKAT PRESS
Port Washington, N. Y./London

JOURNAL OF A RESIDENCE AND TOUR
IN THE REPUBLIC OF MEXICO

First published in 1828
Reissued in 1971 by Kennikat Press
Library of Congress Catalog Card No: 72-130332
ISBN 0-8046-1393-1

Manufactured in the United States of America

TO

JOHN TAYLOR, Esq. F.R.S. F.G.S. &c.

MANAGER OF THE REAL DEL MONTE

AND BOLAÑOS COMPANIES.

My Dear Sir,

In dedicating to you the following Journal of my Tour through several of the Mexican States, I am happy in having it in my power publicly to express my grateful sense of your kindness to me, throughout the official intercourse which took place between us during the period of my engagement in Mexico ; and to subscribe myself,

My Dear Sir,

Your grateful obliged Servant,

G. F. LYON.

THE interest which is felt by the public in all that relates to Mexico, renders me anxious lest the limited quantity of information I am able to communicate should occasion disappointment. It is therefore necessary to state, that the following Journal pretends to nothing more than an account of my personal adventures during a residence of eight months in various parts of that country. Although it does not contain matter of much importance, it is a faithful narrative of what I saw : and I trust it will be found to give a fair representation of the state of the Republic, and to add in some degree to the very small stock of information which exists respecting the people and general appearance of that portion of the New World.

I have further to regret, that my materials, already too scanty, were rendered more so by the loss of many papers, and the greater part of my collections, in the wreck of the Panthea, in which I returned to England.

CONTENTS.

VOL. I.

CHAPTER I.

CHAPTER II.

CHAPTER III.

CHAPTER IV.

CHAPTER V.

CHAPTER VI.

RESIDENCE AND TOUR

IN

MEXICO.

CHAPTER I.

HAVING been appointed one of the Commissioners for the Real del Monte and Bolaños Mining Companies, I left England on the 8th of January 1826, in the Company's brig Perseverance, taking under my charge a large party of artificers intended for the mines of Real del Monte and Bolaños.

My reasons for undertaking an employment of this nature in a time of profound peace, and at a period when no expeditions of enterprize were going forward, cannot be interesting to the public. I therefore avoid detailing them: and although on my outward passage I wrote some few remarks on Madeira and Jamaica, to divert my thoughts from all I had left behind, it would be equally unnecessary for me to publish them; those places being already too well known to the general reader to require further description.

On the 9th of March, having crossed the Gulf of Mexico, we struck soundings near midnight, and on the 10th saw, from the mast head, the coast near Tampico *; indistinct and low, while in thirty fathoms water. Shoaling to twenty-five fathoms, we then saw the loom of the sand-hills from the deck,

* I mention the soundings on approaching the shore, as they may be important to seamen; although I have more fully described the river Panuco and its Bar, in some notes in the Appendix.

at a distance of about eighteen miles; and having in the afternoon distinctly made out the Bar, we ran for it; and when in six fathoms, a pilot launch came out in answer to our signal guns. The pilot at first refused to take so large a vessel * over the bank,— probably to enhance the value of his services; but he was ultimately persuaded to do so: and we safely passed this impediment, which in charts and books of pilotage is described as being so dangerous. On this day, however, the passage was perfectly smooth, and the break of the sea on the shoal points which bounded it, could scarcely be said to exceed a ripple. Nineteen feet and a half was our shoalest cast,—a most unusual depth,—and the Perseverance was said to be the first vessel drawing above twelve feet of water which had ever crossed this continually shifting barrier †.

We had no sooner anchored in the river Panuco, off the little cluster of huts at La Barra, than a crowd of all descriptions of men, women and chil-

* She drew twelve feet and a half.
† See notes on the Bar.—Appendix No. I.

dren hastened on board; and the Custom-house
officer, accompanied by a dirty, ragged, ill-looking
man, who was styled Captain of the Fort, seemed
fully disposed to give us some trouble: but wine
and cakes distributed to their families, and cigars to
themselves, effected wonders; and at sunset I was
permitted to accompany the captain of our vessel to
the town of Pueblo Viejo, at which the Commandant resided. Night soon closed on us, and we
rowed for above two hours, against a strong current, up a stream of half a mile in width. The
sound of our oars aroused the large cranes,
herons, egrets, and innumerable other birds from
their rest, and they fluttered in blind confusion across the surface of the stream; while myriads of fire-flies were flitting amongst the dark
mangroves which dipped their closely woven
branches in the water. Tree-frogs and crickets,
which abound here, almost deafened us with their
shrill thrilling notes; and to add to the delightful
novelty of my first evening in America, we were
hailed in our own language from an invisible boat,

by a gentleman who, suspecting us to be strangers, offered to pilot us to the town. We soon reach‑ ed the house of Mr. Robertson the American Consul, to whom we were consigned, and met with a most kind reception. Seeing that we were tired, hungry, and wet with the heavy night dew, he obligingly supplied all our wants, and provided us with beds in his office;—but sleep was quite out of the question. Dogs, pigs, and restless cocks which began crowing at midnight, would in them‑ selves have been sufficient to banish rest from a stranger; but at about 1 A. M. of the 11th, a storm of rain, thunder and lightning set in with great fury, and in a few minutes actual rivers were rushing through the town. This was a severe " Norther," and we were rejoiced at having entered the river, as, had our vessel anchored outside, it would have obliged her to make an offing, and many days might have elapsed before she could have reached the port again.

In the forenoon I waited on the Commandant, for whom I had dispatches, and then presented to

the Director of the Customs and the Alcalde, the
letters of introduction with which I had been
favoured by D. Vicente Roccafuerte, charge d'affaires for Mexico, in London.

The Commandant obligingly assured me that
"everything he had was at my disposal;" the Alcalde begged me " to believe that he was my servant;" and the chief of the Customs having " kissed
my hand, and entreated that I would lay my commands on him," then proceeded to throw some
impediments in my way.   Mr. Robertson very
kindly hired a windowless room for me in one of
the most respectable houses in the place, the mistress of which was better known by the name of
the Gachupina (a term of reproach applied to
European Spaniards) than by her proper appellation of Doña Francesca.

This lady, who had the reputation of being rich
and cleanly, was quite distressed at not having time
to whiten my room ; but two Indian girls were instantly set to work to wash the earthen floor and
make me comfortable. My landlady was all polite-

ness, and I clearly saw that she entertained no mean opinion of her own good breeding and address.

She was generally allowed to be one of the most respectable ladies of Tampico; and although a certain Don Antonio, who assisted in taking care of her shop and her fair self, was by no means her husband, she piqued herself on her irreproachable character, and the high estimation in which she was held on account of her wealth. Her age might have been about forty-five; her person was fat; and when in her morning costume, which consisted of a shift tied round the waist with a string, and with a cigar in her mouth, her whole figure was particularly attractive. This charming person and I formed an interesting picture every morning at about six o'clock, as she stood leaning over a little wicket which kept the pigs, dogs, cats and poultry from coming out of the yard into my room. While we cosily smoked the cigars with which she favoured me, I drew in lessons of Spanish, by conversing with her, and listening to constantly repeated assurances that she was an " old Spaniard" and a woman of sense, although, in common with nearly all the native ladies of the place, she did not possess the accomplishments of reading and writing.

In the evening, the Commandant and his wife, with some officers of the garrison, paid me a visit at the Consul's house, where his sister amused the party by playing on the piano-forte; and a host of half-naked natives soon crowded round the door and windows, which latter at Tampico have no glass, and seemed highly delighted with the music; some of the children remarking with astonishment, that the Señora " read a book" while playing.

When the visitors retired, I adjourned to my own room, that I might endeavour to sleep,—but it was all to no purpose. At my outer door stood a pig-sty, containing three old sows,—mothers, to my misery, of a numerous progeny, which had been separated from them in order to forage in the odoriferous streets. The moanings and cries in every tone to which the voice of sows can be modulated, were answered with interest by the young pigs; for, being too large to creep through the bars to their parents, they solaced themselves by squealing throughout the night. To add to this, the dogs, of which every house has several, barked without interruption.

March 13.—On rowing down to the brig we saw five or six alligators (which are more generally distinguished here by the names of "Lagarto" or "Cayman") in the water, and others basking in the sun on shore. I landed opposite to one of these in a little creek, and crept cautiously through the high reeds to get a shot at him: when just as I had reached the water's edge and was preparing to fire, a very large one, on which I had nearly trodden, awoke from his slumbers, and plunged into the water, frightening the other sleeper, and not a little startling me, one of these creatures having very recently killed a woman. On returning home I shot a beautiful rose-coloured spoonbill, and an egret, both of which birds abound here.

March 14.—At the new town of Las Tamaulipas (about three miles to the northward of Pueblo Viejo, and in a different State,) I experienced great difficulties with the Custom-house officers, who would do nothing towards clearing the vessel. Their only working hours at any time were from 9 A. M. to 1 P. M., after which period no entreaties could induce them to move. Even in the very small portion of the day set apart for business, if a cock-fight was to be held, if gambling engaged them, or if they were lazy, duty was very quietly deferred until the morrow, when in all probability the same excuse would be again offered; and as in this land of liberty and equality scolding these people would only make them worse, patience is the best policy, and is one of the greatest blessings which a traveller in the Republic of Mexico can possess.

I this day killed an alligator, but he sunk before I could take firm hold of him. One of our party, however, was more fortunate, and procured a small one of six feet in length.

On the 16th of March I dispatched a party of the

steadiest of the people whom I had brought out, to Real del Monte; and on the following morning the remainder of the men and two women followed to the same place. It would be in vain for me to attempt describing my feelings of delight, when I saw the tail of the last mule turned towards me, and knew that I had now got rid of all my charge, except three or four of the best conducted amongst them. Only those who have been shut up for two months with a set of mechanics who imagine themselves too good lawyers to be controlled, can at all enter into my feelings on this occasion. Indeed, when it is considered how different is the confinement on board a vessel and the want of occupation, from their usual habits of life, it is to be expected that the charge of superintendence of these people is attended with much trouble.

At the village of Tampico, seven miles to the southward of Pueblo Viejo, I visited with a friend the church, in which we found the old Padre wandering about in a long blue gown, the counterpart of that which Edie Ochiltree is described to have worn. It being a Saint's day, a number of women were seated on the ground in prayer at the shrines; some of them having a row of lighted candles stuck in the earthen floor in front of them. The Padre, who appeared and spoke as if very tipsy, took us behind the altar to see an image of our Saviour, which, except on particular feast-days, is never exposed to vulgar gaze. The figure was as large as life, ill-proportioned, of a ghastly yellow colour and having indications of the veins painted black. An apron of red damask, garnished with gold tinsel roses and tawdry ornaments, was tied round the waist, and a wig of immensely long hair covered the shoulders. The whole figure reminded me forcibly of the horrid creation of Frankenstein. It is, however, celebrated for the miracles it has performed: and having, according to the priest, been found by some soldiers so immediately after the Conquest as to do away with all probability of human agency, is considered as being sent from Heaven!

March 18.—On this evening the whole town was

in motion. The officers and their band paraded all the streets, and ultimately centred at the dwelling of the Alcalde or Mayor, who kept one of the most reputable drinking-houses in the town, and rather piqued himself on having as few drunken people gambling on his counter, as were to be found in the shops of any of the other Authorities. It happened to be the eve of his fête, when it is customary at this place to offer the compliments of the season. On the following day, his " dia de nacimiento," the officers of the garrison, in their full uniform and preceded by a discordant band of drums and fifes, marched down at eleven to partake of the feast, to which I also was invited. In the centre of the receiving-room was placed a large table, loaded with the most admired dishes of the country. The delicious fish called Sapo (or frog, from its resemblance to a tadpole) lay floating in an ocean of oil, garlic, pepper, raisins, and other savoury articles. Meats in all sorts of shining pies, and in a variety of stews, abounded ; and every little vacancy between the dishes was filled up by a capacious black bottle.

We all stood to partake of the abundant feast " con franqueza ;" and compliments having been plenteously bestowed in all directions, a copy of verses was read by the Commandant, informing us that our little round good-humoured host possess-ed every virtue under the sun. This, with its re-joinder, was the signal to withdraw; and having all wished that the Alcalde " might live a thousand years," we lighted our cigars and took leave.

March 22.—I procured two of the extraordinary fish which are here called Cuttan, and known to the American seamen as the Alligator Gar, or Gar-fish.

The head and jaws are precisely those of the al-ligator, and two of the front teeth of the lower jaw protrude through holes in the upper as with that creature. In each side of both jaws were two rows of very strong sharp-pointed teeth, of which the outer were by far the largest, and the rows were separated by a deep groove. The palate also was grooved and of a very hard enamelled bone, and even the short tongue was thus defended. The whole body was

covered with scales of a particularly hard bony nature, and of such strength as scarcely to yield to a
strong chisel and hammer.   This fish has two ven-
tral, two pectoral, and two anal fins; but no dorsal
fin. The fins are small and nearly of an equal size.
The eyes prominent, with the faculty of turning in
every direction. The fish, which exceeded five feet
in length, was considered as a small one of the species, which I should conceive to be of the Esox, or
Pike genus, from its similarity in some respects to
the *E. osseus.*

On the 23rd, a party of us visited some islands
in the centre of Tampico Lake, at about seven miles
distance from Pueblo Viejo. On rounding the first
small woody islet, we glided from the effects of a
fresh " Norte" or Norther which was blowing, into
a smooth sheltered pool thickly overhung with man-
grove and other trees, on which were sitting hundreds of cranes, egrets, rose-coloured spoonbills,
brown and blue herons, and various other birds, all
as thickly crowded as the tenants of our English
rookeries.   In shooting our specimens, we alarm-

ed two very large alligators, which could not reach the lake but by walking along an extensive sand-bank; and we were thus enabled to observe them in their clumsy progress, which can very rarely be the case, as they are usually found lying so close to the water's edge that they reach it by one spring. From the small island, on which we procured many hatsful of eggs, we proceeded to another, on the north side of which were innumerable nests of cranes and the roseate spoonbill, each containing two or three nearly fledged young ones. The mangroves are so closely woven together at this place, that there is little difficulty in climbing amongst them, and even walking on their tops. We consequently scrambled about to admire the pink clusters of little spoonbills and the pure white down of the young cranes, all as large as barn-door fowls, and offering the most beautiful contrast imaginable to the deep shining green of the leaves amongst which they were sheltered.

Having completed our collection, we returned home, and I carried with me a couple of young spoonbills, which soon grew so tame as to come at my call and follow me wherever I chose: but these beautiful creatures died when put on shipboard for a passage to England.

March 24.—In the afternoon of the 24th we most unexpectedly received information of the arrival off the Bar, of General D. Miguel Barragan, Governor of the State of Vera Cruz, of which Tampico is the northern limit; and it was expected that all the authorities should go in full dress to meet him; the American Consul and I receiving an invitation to join the party. This was one of the oddest water-excursions imaginable: there were about eight boats, of all sorts and sizes, crowded with officers in cocked hats and plumes; some aldermen, with a handkerchief as a substitute for a hat; a few of the shopkeepers in full suits of black, and most motley crews of rowers: the whole party, down to the smallest boat-boy, screamed, roared, and gave advice to the Commandant as to the most imposing order in which we should arrange the squadron to receive the General. He, unfortunately, popped unexpectedly upon us in a short reach of the river, while the boats had formed in a cluster for the purpose of a general ignition of cigars. The great man, however, was happy to avail himself of the flints and steels, and we then returned in state up the stream.

March 25.—I this morning discovered one of the causes of my nightly disturbances; which was, that a large house within a few paces of my door contained no less than twenty game-cocks, each tied by the leg to a short stake planted near a neat mat which served as the roost or bed. These birds were all the property of an alderman, the most scientific and adventurous cock-fighter of the whole corporation.

March 26.—I visited and sounded the Bar on this day *. Several pelicans were fishing on the shoals, and many thousand terns sat clustering in one flock on the northern point. Landing at the little village of La Barra on our return, we found

* See Appendix No. 1.

the General there, and received an invitation to join his party, who presented us with large wine-glasses of poisonous aguardiente.

I was abundantly questioned, and in a few instances with some degree of mistrust and jealousy, as to my motives for examining the Bar, and my opinion whether this formidable barrier admitted of improvement, which was considered as impossible by the whole party. To give them some idea of what skill and labour could accomplish, I described to them the Plymouth Breakwater. When I had done so, a general silence prevailed; and though the company were too polite openly to question my veracity, I could yet plainly perceive that in their hearts they doubted my narrative.

This being Easter-eve, was the first of those days especially set apart for gaming and idleness; and at about 9 o'clock I went to the Plaza (an open space near the church), where I found many hundred people already assembled to amuse themselves. A large circle surrounded by spectators and dancers was expressly set apart for fandangos, which, whatever they may be in Spain, are in the New World much inferior in grace and activity to the common African Negro dances; though the latter, it must be confessed, are usually to the sound of tin pots and empty gourds. Here the music was somewhat better, though not less monotonous; and consisted of a guitar, a rude kind of harp, and a screaming woman with a falsetto voice. Beyond the fandango stood a range of booths, beneath which, men and women of all descriptions, old and young, rich and poor, officers in full uniform and beggars in rags, were gambling with the most intense interest; and individuals who from their appearance might be considered objects of charity, were fearlessly staking dollars,—some even adventuring a handful at a time. The favourite game was that called "Chusa," which is played on a deep saucer-shaped table, and resembles the E. O. of England. All round the Plaza small groups of Indian and other women were seated on the ground with little charcoal fires, at which they occupied themselves in preparing coffee, chocolate, fish, and other eatables;

while under the Chusa tents, spirits of all kinds were sold in profusion.

March 27.—A gentleman at Tamaulipas permitted me to take drawings of two very perfect Mexican idols. They were of basalt, and had with many others been dug up near the spot; but their companions had met with the usual fate of these interesting objects, having been broken up as building materials ! Some European Spaniards who overlooked me as I sketched these figures, could not refrain from showing their amazement at my taking so much trouble about " such ugly things" (*cosas tan feas*),—a remark which soon ceased to astonish me, as subsequently to this period I had frequent opportunities of observing the extreme indifference of the generality of the Spaniards to every thing connected with the history of the aborigines of the country.

When the oppressive glare of the sun had ceased, and the cool evening breeze set in, Doña Francisca announced to me her intention of visiting the Chusa; and inviting me to accompany her, walked

there in great state between D. Antonio and myself, preceded by her three servant-maids, one of
whom was in her Indian dress, and had charge of
the cigars for her mistress. We found our way
to the largest gaming-table; at which Francisca,
having elbowed some ragged women off the only
bench in the place, established herself in full play.
Fine ladies with mock jewels, and women of all
shades and colours, with every variety of men,
crowded thickly round their favourite game; and
my landlady having succeeded in getting the balls
into her own hands, became entirely occupied in
throwing them with such gestures or turns of the
arm as in her opinion would insure success. Before leaving the Plaza, where Francisca remained
playing until nearly daylight, I made my way
through the crowd to take a last peep at her; and
saw a fellow to whom I had paid a real* in the
morning for sweeping before my door, and who was
almost in rags, standing opposite my fair friend,
acting as banker to the table, at which I suppose

* The eighth of a dollar.

he had been successful.   He had squeezed a real
into his ear "para Fortuna," and ventured his dollars at every turn with the most perfect sang froid.
The apparent indifference to losses, and apathy
when successful, is very remarkable with all classes
of Mexicans: but they gamble so incessantly, that
I should conceive all excitement in this dangerous
passion must be deadened, and that the love of play
at last becomes a disorder rather than an amusement. I have frequently seen a couple of poor por-
ters, who had not a farthing of money, sit gravely
down in the dust with a greasy pack of cards, and
anxiously stake their respective stocks of paper cigars, until one or the other became bankrupt.

March 28.—I joined the General and a large
party in a visit to the newly established Lancas-
terian school, held in the neatest building in the
town, which was purposely constructed for it. Although the establishment had only existed two
months, there were already two hundred boys and
about twenty girls on the books, many of them not
more than four years of age.

The school-room for the boys was large and airy, the forms and desks neatly made of a wood resembling mahogany. Round the white-washed walls were painted, in large black letters, various well selected moral sentences; and the whole establishment wore the appearance of great order and regularity. The master (a native of Madrid) read a long paper to the General on the great advantages to be derived from the school; but few of the party seemed to feel much interest about it. Some of the boys then wrote and read; exhibiting proofs of being far better scholars than most of the authorities who sat in judgement upon them.

The master, who was a quiet well-informed man, gave me much interesting information as to his charge, but complained that the principal people of the town afforded him little assistance.

The Bible, in Spanish, is permitted to the children unconditionally; but it appears that few, either of the old or young in Mexico, are inclined to read it. It is indeed to be expected, that amongst a people to whom information on all subjects

has so long been denied, and whose religion is
framed from legends of which European Catholics have never heard,—the Bible in its pure simplicity should not appear to bear the stamp of
truth, and that the few selections from it should
have been warped entirely to the purposes of keeping a naturally tractable and very credulous race
of men in a state of blindness in religious matters,
bordering but too closely on idolatry.

March 29.—I dined with a party of thirty-five at
the house of a native merchant, who gave a " con-
vite" in honour of the General. The eating part
of the entertainment, which occupied above three
hours, was served up in one or two dishes at a time :
but we had scarcely been seated ten minutes, before
a " bomba" was proclaimed; and other toasts followed in rapid succession, generally in verse. It
was extremely amusing to watch the progress of a
sentiment, which I at first imagined was from
the feeling of the moment,—but no such thing :
a gentleman would rise suddenly on his legs,
with an air of the greatest enthusiasm, would

wave his hand above his head with a joyous shout, and proclaim a "bomba!" All the party would then rise to second this animating burst; while bumpers were filled and silence was obtained: after this, the proposer of the toast would very gravely produce from his waistcoat pocket a ready-made copy of verses, which not being of his own composing he could not very easily read; and then, all having loudly cheered this genuine son of Anacreon, the company would sit down until a similarly bright and equally original sally was made by another of the party. Unfortunately, on this festive occasion all the verses, and there were at least thirty, turned on the same subject: pompous, overstrained compliments to the "Invincible" the "Immortal" "The Hero of the Age, General Barragan;" or, "To the Success of the Mexican Republic, the Envy and Admiration of the World!"—I was most heartily rejoiced when, at 8 o'clock, all this ceremony was over. We were then enabled to go, in good time, to a ball at the drinking-shop of the cock-fighting alderman, who did not, however, honour us by his presence. I know not what music we had at the ball,—the performers, certainly not of Colinet's band, being hidden in a crowd of Indians and idle people who stood in the court-yard.

The ladies, who at Tampico had not at this time arrived at the refinements of any kind of stays or gloves, sat in a stiff formal line at one end of the room, all smoking very seriously, while at the other the gentlemen also solaced themselves with the same ball-room luxury. The General, who had been much in the more polished parts of Mexico, looked rather bewildered on being pressed to dance; but complied with the wishes of the company, and swam through the labyrinths of a Spanish dance with a most woe-begone countenance.

Following his great example, I did the same; but with no great success,—my partner, a very active shrivelled little old woman, being so provoked at my inability to exhibit her graces to advantage, as very unceremoniously to leave me standing in the middle of the room, while she retired in high wrath to a corner, and sought consolation in a cigar.

March 30.—A gentleman at Las Tamaulipas very courteously presented me with several small figures and imitations of Indian weapons carved in bone: these, with a skeleton, had recently been dug up in sinking the foundation for a house in this new town, which, as I have elsewhere remarked, is evidently on the site of one of the Indian cities discovered here by Juan de Grijalva. I subsequently received from the same gentleman a little idol in terra cotta, which was found on the day he brought it, in digging a well. It was similar to others which I procured in Panuco, and the face and head-dress in all the specimens resembled each other in an extraordinary degree.

Having dined with the Commandant, and seen the poor modest General sadly persecuted with compliments, I accompanied him and our entertainer to the Chusa, at which the Commandant (a colonel in the army and governor of Tampico)

soon found his way to the table where his lady
and my friend Francisca sat playing with great
animation. In a moment this high official character was deeply engaged in betting and staking his
dollars with all the ragamuffins whom I have
already described as frequenting the Chusa. The
boatmen and cargadores rather predominated at
his table; but they ventured deeply against him,
and in so doing lost much of their money. The
General, meanwhile, rambled about in the crowd
in a large gold-laced hat lined with black ostrich
feathers, bearing in his hand a tumbler of aguardiente, which he offered to all his acquaintance,
to whom he also presented the cigars he had purchased from time to time at the booths. The
great familiarity of manners amongst the Mexicans may appear to strangers as the effects of the
recent changes and assumption of a republican
form of government; while in fact it is adopted
from the custom of Old Spain, where the domes-
tics, bred up from generation to generation in the
same house, inherit all the privileges of familiar

intercourse. With this foundation, society in New Spain is far more debased than in the mother-country, owing to the cruel state of ignorance in which it has been the policy of Spain to keep its Transatlantic subjects. It may be plainly perceived, therefore, that the equality of education, breeding, and knowledge of the world, have brought the beggar and the noble on joking terms together. Things must now rapidly improve: and when the women are allowed their proper station in society; when the female children are restrained from playing in the streets, or with the dirty persons who act in the capacity of cooks; and when stays and ablutions are introduced, and cigars forsaken by the softer sex, the manners of the men will be materially changed. In the southern provinces, the effects of free intercourse with Europe have altered not only the forms of society and dress, but have given to the manners of the better classes a degree of polish which will speedily be imitated by their more rustic neighbours.

General Barragan was a mild obliging man, who, instead of proffering attentions according to the established forms, did all in his power to serve his friends; and he afforded me assistance whenever I required it, with a cordiality which is not often to be met with in Mexico.

April 1.—Having learnt by a letter from a party who had been sent forward to Zacatecas, of their detention by the Alcalde of Altamira, I rode there to arrange matters; and the impediments arising from a little overstretch of power being removed, the people had permission to proceed forward on the morrow. When this business was settled, the Alcalde, who owned the only little public-house in the place, was very attentive, and took me to the top of the church, of which he is the head chorister. Hence I obtained a view of the Lake, and the neglected, forsaken town of Altamira,—once so crowded and of such commercial importance, but now containing scarcely more than one hundred souls, the inhabitants having removed to the new and rapidly rising town of Tamaulipas.

April 2.—It was a delicious morning when I left Altamira; and the fresh dew, as it rapidly evaporated under the first rays of the morning sun, gave out all the sweets it had imbibed from the wilderness of flowers. Nothing indeed could exceed the fragrance exhaling from the woods for the first two hours of the day. Its fineness had called forth some beautiful varieties of small snakes, many of which, of an exquisitely brilliant green colour, were gliding about the pathways. I now saw the smaller Mexican pheasant, and the cries of the chachalaca resounded from every part of the forest.

The distance between Tampico and Altamira is about seven leagues. In the afternoon I received a visit from a very extraordinary posture-master, a little fellow of about forty years of age, with a countenance alarmingly like that of a baboon. Enormous ears stood at right angles with his head, which was thinly covered with single crisp brown hairs: his figure was almost deformed, yet his attitudes were certainly equal to those exhibited by Mazurier.

A Mexican man-of-war brig and two schooners anchored off the Bar in the afternoon, when I had an opportunity of seeing some of the officers of the incipient navy of the Republic.

April 3.—On crossing for Tamaulipas I shot the largest alligator I had yet seen, and which could not have measured less than twelve feet. On being struck, the blood spouted copiously from its wound, and the air became strongly impregnated with a pungent odour of musk. As this creature slid from the bank into the water while struggling in death, I saw no more of it until the following day, when its carcase, half bared by the vultures, was lying on a sand-bank, and in a state too offensive to be approached. I lamented not having obtained the skin of this creature, as it was of a bluish black colour; while the usual hue of the Tampico cayman is a brownish green.

VOL. I.

April 5.—In an expedition down to the Bar, I paid a visit to the Castillo, or Fort, which is placed at its southern side. It is a small space rudely surrounded by upright stakes planted in the sand: their height is about that of a man. Vacancies are left on a level with the ground; and five guns, of as many different calibres, frowned through these embrasures. First appeared an iron six-pounder carronade on a ship-carriage; next was a long twelve, with a similar mounting; by its side stood a small brass field-piece on high wheels; and last of all, a venerable nine-pounder showed its honey-combed muzzle. Of all these engines of war, the brass swivel was the only one not choked with rust: yet the natives consider the Castillo as a "place of proof;" and it was originally constructed for the purpose of intimidating a French squadron which some time since was expected to appear on the coast to assist the Royalists!

When I peeped over the stakes at the fortress, the officer of the guard rose hastily in his blanket, cried " Vigilancia" in a tone of thunder to a slum-

bering Indian soldier who was lolling in the burning sunshine, and then proceeded to inform me
that no strangers were permitted to look into the
castle. I could not avoid assuring him that I was
not going to take any hints from it; and then at a
respectful distance made a sketch of this very curious place,—not, however, without interruption ;
an Indian soldier having somewhat peremptorily
notified to me, that no one was allowed to write a
" descripcion " of the Fortaleza, without permission of the government.

April 7.—Amongst my other duties I attended
at the Custom-house at Tamaulipas to pass ten
thousand dollars which we had brought with us;
but learnt with astonishment that no money
coined abroad could be landed ! Dollars of every
part of Southern America are prohibited, or at
all events considered as not proper to be introduced; and even the Spanish pillar-dollar is
objected to. Our agent and myself were required to enter into a bond relative to this money,
" that if at any *future period* the government

should impose a duty on the importation of dollars, we should be liable to pay it." In consequence, however, of this singular clause, we entered the money as landed for exportation, sold it to advantage, and it was re-embarked by the purchaser without paying the three per cent exportation duty, which would otherwise have been due to the state. I mention this as one of the many impolitic laws relative to the public revenue, which is further prejudiced by the import duties on every article of commerce, amounting at this period to about thirty-five per cent, exclusive of still further Internacion duty to the amount of twenty-five more.

The consequence of this measure, has been the introduction of a regulated system of smuggling; for conniving at which, each class of officers of the Customs had its stated fees: by these douceurs the purposes of the merchants were fully answered, and at least half of the established duties saved to them, while but a small proportion of benefit accrued to the revenue of the state *.

* An administrador of the Customs, whose salary was one

April 9.—We rode this evening to the Mira, or Look-out, which is on a thickly wooded hill at the back of Pueblo Viejo, and commands a view of the sea and outer surf of the Bar. A delightful view of inland scenery,—a portion of the River Panuco, and the wide-extending Lake of Tampico,—is also obtained; while the New and Old towns, with the vessels at anchor off Tamaulipas, form a coup d'œil, which of its kind is unequalled in any part of Mexico. There are two small bamboo huts on the eastern side of the hill inhabited by Indian families, who support themselves by the sale of a beverage expressed from the sugar-canes, of which they have a small plantation. It is the custom of all classes at Pueblo Viejo to make their Sunday evening's lounge to the Mira and drink this liquor, which, though very inferior to Pulque *, is in great request. On this day the beautiful wood was much

thousand two hundred dollars per annum, retired at this period, after about eight months official residence, with one hundred thousand dollars !

* This is only procured in the more temperate regions of the table land.

crowded with pedestrians, winding amongst the intricate and shady mazes of the thicket. Indians with their families, half-casts, negroes, and creoles, in great variety of gay costumes, rendered the picture extremely pleasing. Here and there the tinkling of the small Indian guitar, or joyous songs by parties who carried fruit and provisions for their evening feast under the trees, gave an indescribably lively character to the scene, heightened as it was by the large flights of bright green parrots and parroquets, the restless cardinals and crimson-crested woodpeckers, with many other beautiful varieties of birds, all in happy activity as the burning sun retired.

April 11.—After having lived for a length of time in noise and misery at Pueblo Viejo, I at length succeeded in lodging myself at Tamaulipas, in a large white-washed barn, consisting of one room; for which, without a single article of furniture, I paid two dollars and a half per diem, and was congratulated on having found so cheap a lodging in the New Town. I am not intending to be fastidious, neither am I one who complains of lodging, diet, or other inconveniences to be met with in the ordinary course of things; yet anything was happiness to me after the constant noises at Francisca's: and I noted with great satisfaction the first entire night's rest which I enjoyed uninterrupted by dogs or cocks since arriving in the country. Hogarth's " Enraged Musician" never suffered more than I had done, in consequence of the utter impossibility of enjoying one quiet hour to attend to my business. I may be pardoned for giving some description of my troubles, in order to show that, with every disposition to accommodate myself to circumstances, it was quite impossible to sit easy under such a constant din. The back door of my room, which supplied the place of a window, opened into a yard in which eighteen hens and numerous chickens were wont to ramble. A horse tied to a tree neighed at intervals to its two responsive fellow-servants in a small open stable. Four dogs, of various voices and most provoking tempers, growled and barked constantly at each other, or at five starvling cats and their kittens, from daylight until sunset, and from dark again until morning. The projecting eaves of the house formed a kind of covered way, to protect Doña Francisca from the sun,—sheltering also one of my greatest torments. On a pole, suspended by a rope at each end, swung a favourite parrot, which talked incessantly, and very much in the same key as its mistress. Patches of Spanish songs, terms of insult and opprobrium, expressions of endearment, curses, and orisons to the Virgin, succeeded each other in a rapidity of utterance altogether peculiar to the mistress and the bird. It was also my fate to hear two young Indian women and a little talkative girl grind maize, slap tortillas, sing, gossip and laugh abundantly, within two yards of my door; and at meal-times, the smell and sound of fish and other viands frying in oil was added to the harmony. Francisca, of whose tones words can give no idea, had adopted a sickly, yellow, and fretful little child, with a perpetual scream and constant restlessness. The squallings of this poor infant, which were faithfully imitated by the parrot, elicited alternately, coaxings, scoldings and whippings from its madrina; and the general result was, that the wretched baby on being turned as a punishment into the yard, would creep in and sob in one corner of my little room. I say nothing of the pigs and a kid or two, which, unhappily for me, had excellent lungs, or of Don Antonio's constant squabbles about politics with a sickly fellow-lodger of mine, who, when not otherwise occupied, dropped in to tell me of all his ailments. But when I add to this list of miseries the insufferably powerful fangs of the fleas, bugs, mosquitos, sand-flies and garrapatos *, which feasted upon me in those moments when I most required rest, my situation may be easily imagined.

On the morning of the 14th of April I left Las Tamaulipas with my servant Marriot and one of the Company's men, in a fine large canoe formed

* A small kind of tick, which in great numbers bury their heads in the skin and are very difficult of extraction.

of a single tree, and proceeded on the examination
of the River Panuco, of which so little has been
hitherto known. We started with a light sea-
breeze up this beautiful stream, and our general
rate of progress through the water was about
two miles and a half in the hour. Our two canoe-
men rarely paddled the boat, or entered the centre
of the stream, unless when sailing; but at all other
times pushed it forward by putting their oars to
the bottom, and keeping for that purpose close to
the bank. A portion of the stern part of our canoe
was covered with an odd kind of awning, to shelter us from the broiling sun by day and the heavy
dews of the night. This was composed of a rough
frame of green sticks, bent from side to side of the
boat, and covered with three fresh bullocks' hides,
the smell of which while drying over our heads was
particularly offensive; they afterwards became stiff
and semi-transparent, and the sun's rays penetrated
to such a degree as to make us fancy ourselves exotic
plants undergoing the process of forcing. To add
to our discomfort, the awning was so low as not

to admit of our sitting upright under it. The banks on either side the river were thickly and luxuriantly covered with small trees and shrubs; and at the end of about twelve miles from our outset we came to the commencement of the Rancho de San Pedro. The name " Rancho " is in this country applied to the large pasture districts in which horses and cattle roam almost wild; and when they are wanted, either for the market, for taming, or other purposes, they are driven into inclosures, or caught by mounted horsemen with the Laso.

This process is already too well described by Captain Basil Hall for me to say more about it. The people who attend to these Ranchos are called " Rancheros " and " Vaqueros," and are a fine, active, athletic race of men; much more simple and well-mannered than those who live amongst the busy world. A Rancho " de Ganado" is a cattle farm; " de Cavallada" or " Mulada," for horses or mules; and " de Ganado menor," for sheep and goats. " Haciendas" may be more properly called immense farms: grain is cultivated on them; but at the same time the rearing of cattle in great abundance is not neglected. The lands are generally surrounded by stone walls: a kind of village is established round the granaries and dwelling of the owner or his administrador; and every Hacienda is obliged by law to maintain a church on the estate. On the low lands towards the Tampico coast there are few or no Haciendas, and Ranchos alone are to be found near the river.

Passing for some time the banks of San Pedro, we came to the Estero de Chila, another extensive rancho, the cattle of which were either grazing or lying under the shade of the trees close to the water's edge. On this estate, at about three or four miles from the river, is a large lake, from whence I understand that the petroleum which is brought in great quantities to Tampico is collected. It is here called Chapopote, and is said to bubble from the bottom of the lake, and float in great quantities on the surface. That which I saw at different times was hard and of good appearance, and was used as a varnish, or for covering the bottoms of canoes; the general price was four reals (half a dollar) for a quintal (100 pounds). We had seen at least thirty alligators in this day's journey; and off Chila passed three large turtles, which could not have weighed less than two hundred weight each. To our great regret these creatures were awake, so that we were unable to strike them. We also saw one solitary wild turkey; and shot a few birds, of a sort quite new to me. At sunset we landed, and pitched our tent on a bank near the Rancho del Caracol (or snail), where we could purchase nothing but a leathery cheese and some garlic. Inland of our resting-place lay a wide-extended plain, having a few huts prettily situated under some distant trees, and several lakes, round which numerous cattle were feeding. A number of horsemen were distributed about the plain, collecting the stragglers into groups for the night, and whirling their unerring lasos round the heads of the animals, to enforce obedience.

Myriads of mosquitos kept us awake the whole night, and we rose at dawn of day, April 15, but the herdsmen were already out before us scouring the plains in chase of the milch cows, which they drove at a full gallop into folds near us, while women were waiting to tie their hind legs and milk them: after this they were set at liberty till the following day. The cows in Mexico are milked only once in the twenty-four hours; and from the wandering life they lead, and the little trouble taken about them, yield but a small proportion of milk, and that of very inferior quality. Soon after pursuing our route, and just as the sun had burst forth in all its power, we met a canoe laden with oranges; and I gladly purchased a supply of this refreshing fruit, which is brought down the river from Tamasinchate, a voyage of four or five days for canoes descending with the current. Before noon, the sea-breeze, the greatest imaginable luxury in the Tierra Caliente, set very strongly up the river, and we sailed delightfully before it, discovering as we advanced fresh beauties at every turn of the stream. The varieties of new and magnificent trees, covered with the most luxuriant and brilliant parasitical plants, dipping their branches in the current; withering trunks clothed with a verdure not their own, but which flourished on their decay; and the immense up-rooted timber lying grounded in the shoaler parts of the stream, and causing strong eddies amongst their shattered branches,—gave a character to the scene around which to me was altogether new and enchanting. Here we saw the hanging-nests of the calandria and many bright-plumed birds. Lime and lemon-trees, bearing at the same time fruit and flowers, hung most invitingly over the water, and afforded us abundance of refreshing lemonade. In some places, immense willows threw their cool shade over smooth banks, resembling very closely the park scenery on the borders of the Thames; while groups of cattle grazing or sleeping beneath thin spreading branches, rendered these particular views so like home, that it was fortunate we had some other objects to remind us how far we were removed from it. Here an enormous alligator would plunge into the river from his broken sleep on the sunny bank; or a delicate white heron would rise alarmed on the wing, and soar above our heads when affrighted from her retreat amongst the rushes. We saw also on this day a manati, or sea-cow, but it was out of the reach of our shot; and I killed a water-snake as thick as my wrist, while it lay sleeping in the sun on a branch of a decayed tree. To add to the picturesque of our evening scenery, we came up at sunset with an American schooner, which lay becalmed in one of the short reaches while on her way to Panuco for a cargo of fustic (dye-wood). We were at this time abreast of the Rancho del Aguacate, near which the cultivation of maize, the only grain of this part of Mexico, appeared to commence; and as night closed in we passed several Indian huts surrounded by it. We sailed slowly on with a light breeze, near banks quite illuminated by the fire-flies, and the wailing and cries of the solitary night birds gave a peculiar solemnity to the evening; when our ears were suddenly enlivened by the merry sounds of a fiddle and a guitar, proceeding from a small canoe, which glided swiftly past us, and was carrying this little band to a fandango about to be given at one of the Ranchero's huts.

April 16.—We slept a few hours this night in our canoe, which was hung to the bushes at the bank of Topila. At twilight we again moved forward: the morning was cool and hazy, the thermometer being only at 70°; and when the day cleared up, we landed at a Rancho, in a long reach of seven miles, called Torno Paciencia, where we breakfasted on milk. The banks on this day's journey had become very steep and sandy, and the small timber began to give place entirely to very large trees. In the forenoon we passed a small island rich in Indian corn, standing seven or eight feet in height, with pompions and water-melons, some of which we purchased. I afterwards learnt, when too late to take advantage of it, that a very ancient statue was to be seen at this place. Opposite the island is an establishment, formed three years since by some Americans as a distillery of Aguardiente from the sugar-cane. It was at this period at work.　Near this we saw two manati at a distance in the river.　The thermometer at noon in the shade was 83°, and in the sun 94°.　At 8 p.m. we rounded a point thickly planted with maize, and came in sight of the village of Panuco, situated on a high bank, the prettiest spot I had then seen in the country.　On approaching the landing-place we nearly ran the canoe over a naked woman, who was standing up to her neck in the river.　I afterwards learnt that many of the brown sex of this place were excellent swimmers, and that all the natives bathed without the least dread of alligators; these creatures for some unaccountable reason seldom appearing here, although they abound lower down, and are also in great numbers higher up the stream.

While unloading the canoe, two of my friends, who had ridden from Tampico (about forty-five miles) to pass the day with me, arrived, and I procured with great difficulty two small dark rooms in a mud hut. Our search for food and firing then commenced; but neither meat nor wood could we procure for three hours, and we were still longer in obtaining forage for the horses of my visitors. All the town seemed wrapt in a torpid kind of sleep. Money, scolding, and entreaty,—all were tried in turn; but nothing could induce a single soul to assist us. At length, to the honour of the sex be it spoken, a Campeachy female, who had recently awakened from her long Siesta, and had opened her shop door, called me to her, and offered wood, and words of consolation, both of which were gratefully accepted; a few yards of Tasajo, or jerked beef, being produced at the moment, a warm supper was cooked for us. In the mean time I waited on Don Fernando de San Pedro, a very great man, to whom I had letters of introduction, in each of which was particularly mentioned my anxiety to see any objects of interest which might exist in this place, such as Indian antiquities or natural productions of the country. I had been informed by several people at Tampico that a great number of very curious idols were to be found on this person's estate, though the owner of these treasures appeared himself quite ignorant of his riches, and scarcely knew what was meant by " Antiquities," still less by the term " Idol." He was however all graciousness, and permitted me to roam about and make whatever discoveries I might wish. One of my canoe-men was of far more assistance in my research; and his first prize was an odd grotesque-looking figure in terra cotta, used as a child's plaything, for which I was to pay a quarter of a dollar.

Having thoroughly fatigued ourselves by the ramble through the town, we were glad to retire to rest, and still more pleased to learn that there were no mosquitos in Panuco; and I lay down in peace, for the first time since landing in Mexico.

April 17.—My visitors left me early to return to Tampico, and after breakfast I carried a letter of introduction to the Cura, who received me most kindly, and was the means of my passing a very agreeable day. This worthy priest was an intelligent lively man, of about fifty years of age, liberal in his ideas, and ready to give me every as-

sistance and information in his power. His views
respecting the state of the country as to society and
morals, and the necessity of educating the poor,
were just and reasonable, and he assured me that
I should find many of the secular clergy equally
zealous in the cause of improvement as himself.
He laid before me the last census of his little cure,
and told me that any other papers relative to the
Indians, or the "Gente de Razon," were quite
at my service. With this kind guide I again visited Don Fernando, in whose house itself the Padre soon found a curious idol for me to copy. I
had no sooner commenced my drawing, than it excited so much astonishment that half-a-dozen grotesque figures and vases were quickly brought to
me; Don Fernando himself presenting me with a
little bird-shaped whistle of earthenware, having
two holes on each side, so that a kind of tune might
be produced from it. I now found full occupation
for the day, and a whole group of children were
sent out in search of toys, which I agreed to purchase at a medio (three-pence each). In addition

to my acquisitions of this kind, I obtained permission to copy many others, which the owners valued too highly as playthings for their children to part with. The streets of Panuco are to this day thickly strewed with the remains of ancient crockery; and often, after heavy rains, entire vessels and toys are found washed down the water-courses.

In the evening of this day, after the hour of Siesta, I sought out my friend the Padre, who it appeared had been busy in my service, and he gave me three most curious little figures. He also took me to examine a very perfect earthen flute; but the boy to whom it belonged could not be persuaded to part with it. We walked afterwards to see the remains of what the Padre informed me were once Pyramids, and to which the name " Cue*" is still applied, although they are now nothing more than five or six mounds of earth, of thirty or forty feet in height. They lie to the westward of the town,

* This was the term by which the Pyramids were distinguished at the time of the Conquest.—Vide Bernal Diaz, Clavigero, and others.

near each other, and on the plain around them I found several pieces of obsidian arrow-heads, which must have been brought from a great distance* by the warriors who once peopled the banks of the river.

Hence we proceeded through the thickets to several small muddy ponds, where the exhalations are very unhealthy; and although the labour of a few days would suffice to fill up all these places, such is the indolence of the natives that they prefer the annual attacks of fevers and agues, to the trouble of endeavouring to remove their existing causes. I here, for the first time, saw the fruit of the wild pine—Piña del Cardon, or Basuchi,—which is found in a cluster of light yellow plum-shaped fruits, with thick skins, and each containing a small fresh juicy pulp, having a number of seeds mixed with it. This if eaten in any quantity excoriates the mouth very painfully, but is deemed a whole-

* From Pelados near Real del Monte, where there were mines, or rather pits, sunk on the vein of this mineral; or from Cinapecuaro, a day's journey from Valladolid.

some febrifuge when mingled with water. Having rambled through the outskirts of the village, where I picked up many more fragments of obsidian, we went to see a very gigantic Banyan, or Indian Fig-tree, which, situated on a high bank, had thrown down some of its monstrous trunks to the edge of the river beneath; and vessels in need of careening always " hove down" to them. Near this fine specimen were some parched ill-cultivated gardens, where the plantains alone were flourishing. The owner of one of the gardens most graciously presented me with four turnips and one long leafy cabbage, all considered as great rarities, and he also treated the Padre and myself to orangeade and cigars.

At my return home, after dark, I found one of my boat-men in sad tribulation. He informed me that people were coming to imprison him, until he should pay one hundred dollars to the widow of a man whom he had stabbed and killed some months before in a quarrel.

To Englishmen this may seem a very slight pe-

nalty, but it is a far heavier one than is usually in-
flicted even on the most cruel murderers in the
northern states; the general punishment at this
period being confinement, or sentence of confinement for a few days, from which in many instances
the delinquent is permitted to escape and perpe-
trate fresh enormities.

April 18.—This was one of the hottest days* I
had at this time experienced in the Tierra Caliente, and I was happy in any excuse to stay at home
and copy antiquities. The Padre sent me as a
present, several curious ancient toys and whistles,
with one small terra cotta vase very beautifully
carved with those peculiar flourishes introduced in
the Mexican manuscripts. I was also fortunate
enough to procure an antique flute of a very compact red clay, which had once been polished and
painted. It had four holes, and the mouth part was
in the form of a grotesque head.

In the evening, while on my way to thank the
Cura for his attentions, I met a party of soldiers

* Thermometer 8 A. M. 80° ; Noon 86° ; In the sun 104°.

conducting several young men to join the troops at Tampico. They had been forcibly carried away from a fandango, (which by the way is an excellent place for judging of the activity and capabilities of a youth,) and lodged in the common prison until a sufficient number of subjects had been collected. There were fourteen of these unwilling recruits, whom I saw carried off in two canoes so very small that there was but just room for them and their guards to squat down in the bottom without upsetting the boats. Not one of them had a change of clothes or a blanket to cover him, although going a two-days journey down the river, in this crowded state, and in the commencement of the rainy season. Until reaching the water-side they were tied together by ropes, like led horses going to a fair: and of the crowd who witnessed their embarkation, I was in all probability the only person who considered it an extraordinary spectacle, in a time of perfect peace and in a country frantic on the subject of liberty and equality.

April 19.—It rained and thundered, as is usual

in this part of the world, tremendously all night, and at daylight of the 19th we left Panuco to proceed on our examination of the river. Having frequently seen the priest and the best-informed men of the place, I had been enabled to collect something of its history; and on this day knew not a better method of employing my time under a meridian heat, than by retiring beneath the canoe covering of bullocks' hides, and putting on paper the little information I had obtained.

The Panuco, which Cortes conquered with so much expense of lives and treasure*, must unquestionably have stood on the spot occupied by the present town. At the period of the conquest of Mexico it was a place of such importance, that the great captain petitioned Charles the Fifth to add its government to that of New Spain, or Mexico, of which it was, and ever had been, independent. Obtaining his desire, a garrison was placed in the chief town, which was named " San Esteban del Puerto†," an appellation now no longer existing,

* Bernal Diaz.   † Ibid.

although Saint Stephen is still the patron saint,
and the ancient name is the only one at present
used. The descendants of the warlike people who
formerly inhabited the " numerous populous towns
on the banks of the river"* yet dwell in the neighbourhood, but in very diminished numbers. In
their mild dejected countenances no trace is seen
of their being the offspring of those warriors who
defeated Grijalva their first discoverer, Garay, and
the troops of Cortes, who did not effect their sub-
jugation without great loss of men, and at an expense of 60,000 dollars †, an enormous sum in
those days. Time and the tremendous periodical
rains have been insufficient to destroy all vestiges
of the Guastecas nation. The remains of the
pyramids, the quantities of obsidian weapons, the
idols, and the utensils, toys and ornaments in finely
worked clay,—all combine to show that the arts
once flourished to a very considerable extent on
this now thinly peopled spot. Some of the vases
yet retain their colours and vitreous glazing, and

* B. Diaz.   † Ibid. vol. iv. p. 30.

many are of an earth as light and well baked as that of Tuscany ;—while the figures, from their singular attitudes and grotesque expression, might serve as models to the toy-makers of the present day. The flutes, single and double, with two, three or four holes, the oddly shaped pipes and whistles, and the jars modelled into birds, toads, and other animals—all in Terra cotta *, exhibit as much humour as ingenuity, and are found either entire or broken, in such quantities as to induce a belief that Panuco was actually a mart for crockery-ware. I learnt also that at a Rancho, called Calondras, about nine leagues from the town, some very interesting objects of antiquity are to be met with, situated on the side of a hill covered with wild pines. The principal of these is a large oven-like chamber, on the floor of which a great number of the flat stones, similar to those still used by women in grinding maize, were found, and can even now be procured. It is only in the month of May that this place is accessible; as the pines being dry,

* Loza de Barro.

may then be burned from the face of the hill.   It is conjectured that these stones, with a quantity of other imperishable articles of household furniture long since removed, had been deposited in the cave on some flight of the Indians, as being too heavy for further removal.

There still exist at Panuco two Indian " co-munidades*," amongst whom the Guastec lan-guage, to the almost total exclusion of the Spanish, is spoken.   These poor people live unmixed with the whites, who amount to 1500 persons, and who may be called the fixed population.   During the unhealthy months many families come here from Tampico; and in the dry season Panuco is a kind of watering-place, to which people resort for the

		Families.		Souls.
* These are Tānsalichōk, containing	.	138	=	525
Tānquinām		78	=	283
At Tanjuco, nine leagues from Panuco, and appertaining to its Cure, are also }		30	=	101
		246	=	909

Making a total pure Indian population of 909 : a fearful dimi-nution from a nation which according to report once amount-ed to 100,000.

purpose of bathing, the river here being more free
from Caymans than at any other part. For such
families as choose to devote a little trouble and expense to decency, small spaces are staked off near
the banks, and lightly covered with palm branches:
but such niceties are not much attended to; both
sexes bathe without scruple at the same time,
and many of the young women swim extremely
well.

The town is situated on the southern bank of
the river, and was at this time of year at an
elevation of thirty feet above it; but in a more
advanced period of the rainy season, which had
now commenced, the waters frequently inundate the
streets; and it has more than once happened, that
canoes have plied there. Many of the houses are
comparatively good, but by far the greater proportion are of split bamboo, plastered with mud, and
thatched with the fan palm, which is also the covering of the best buildings. There is neither a
school nor any other public establishment in the
town.

It would perhaps be difficult even in this universally lethargic country, to find a more listless, idle set of half-sleepy people than those of Panuco, who for the greater part are Creoles. Surrounded by a soil capable of the highest cultivation, living near a river absolutely swarming with the finest fish, they have scarcely a vegetable, and rarely any other food than Tortillas of maize, and occasionally a lump of Tasajo or jerked beef. The Siesta appears to consume half the day, and even speaking is an effort to this lazy race. Such as are obliged to labour in order to save themselves from starving, obtain their livelihood by cutting dye-woods to freight the vessels which occasionally come up the river for a cargo. These woods are the Moral or Fustic, which sells at four reals the quintal. Sarsaparilla at two reals the aroba, and a wood called Palo Azul, or Blue Wood, which has lately been introduced as an article of commerce, and according to its chemical treatment yields three or four fine tints. All these are brought in from the surrounding forests, yet firewood and charcoal can scarcely be procured in the town. The latter is sold at an exorbitant price, owing to the want of energy in the natives, who prefer receiving it by an eighty miles water conveyance (from Tampico), rather than burning it themselves within fifty yards from their own town.

There are two churches at Panuco, of which the largest is an immense thatched barn-like building, possessing the merit of being so arranged as only to exhibit one Saint: the Redeemer, however, is excluded with those which are rejected ! This is a plan of the Cura, who has determined that the new cedar altar-piece shall only contain a niche for the patron, San Esteban, who from the time of the conquest had been the especial protector of the town. The church has a large and a small bell, placed under a little shed in its front; and on the opposite side of the Plaza stands the Capilla de la Virgen de Dolores, which was founded, and is yet supported, by a body of pious women, who style themselves "The Sisters of Dolores." They are secular, and form a kind of female corporation.

Each sister pays an annual subscription of thirteen reals, for which at her decease she has the privilege of being buried in the chapel; and her immediate relations receive for her either a shroud or twelve dollars and a half, with two dozen wax candles: the former of these the priest secures for the benefit of the departed soul, and the latter are burned round the body.

Panuco is considered a very healthy place in comparison with the other towns on the low lands; yet the fever and vomito occasionally make their appearance, although with diminished violence. The temperature is generally said to be very equal, but hail has sometimes fallen as late as the month of March. The extreme heat and drought of the dry season are effectual bars to the cultivation of wheat and barley; but maize and rice (although I saw none of the latter) flourish on the banks of the river. Beans, chilis (a coarse kind of pea), pompions, with a few sweet and water melons, form I believe the whole list of cultivated vegetables, with the exception of the rare cabbages and turnips.

which I saw in the private gardens. No fruits are reared for sale; but the woods are said to abound in several delicious varieties, which at this season were not ripe. The little comforts of life are here very expensive. Wheaten cakes, equalling our halfpenny rolls, are a medio (three pence) each; onions of the bigness of a finger one penny; and the only cheap article is the Tasajo or strips of dried beef, of which three or four yards sell for a medio.

While I was transcribing these few notes under the frying and not over-fragrant awning of the canoe, the sky darkened; the air became damp and sultry as a vapour bath; myriads of mosquitos flew under our covering: and as we were led to expect a heavy storm, we tied our canoe to the bough of a mangrove-tree, and awaited its coming. I had never before been actually exposed to a regular Mexican shower; and this soon most effectually cooled my curiosity, as it far exceeded every thing I could have expected. Tremendous thunder and very vivid lightning burst from the dense black clouds; a heavy squall swept across the forest, as though it would level the largest trees; and tearing off quantities of branches in its progress, half covered us with the ruin it had caused. The rain now came down in torrents; the canoe was nearly filled, and every thing was set afloat; while as we sat shivering up to our knees in the water, the swarms of mosquitos, ungrateful for the shelter which the awning afforded them, stung us almost to madness.

When enabled to advance again, I continued taking soundings and bearings of the river till we came to a low sandy shoal crossing the whole stream, and having but four feet of water on its deepest part. These low soundings, which continued about half a mile, then deepened to three fathoms. This bar is five miles above Panuco, and is an effectual impediment to the further advance of vessels of burthen. Being a hard sand, it is more than probable that it never very materially alters its position; and in the rainy season, if any vessel had the hardihood to attempt the passage, there would be found an increased depth of twenty or

thirty feet : but it would be a hazardous and fool-
hardy attempt, and its accomplishment could have
no beneficial results. Panuco, therefore, may be
considered as the highest navigable point for any
vessel but canoes.

April 20.—To my great joy the sun burst
through the gloomy black clouds which had again
threatened us; and as the thin gray vapours of the
morning vanished beneath its rays, we had the
promise of a fine, that is to say, a scorching day.
Walking on the flowery bank by the river side,
under the shade of the overhanging trees, the canoe sailed abreast of us before a light breeze ; and
we shot some Cojolites,—dark-coloured birds, of
the gallinaceous tribe, nearly equalling a hen turkey in size, and very delicate eating. We purchased
a large kettle of milk at an Indian hut, and then
pushed on for the pretty Rancho of Miradores, situated near a ferry called the " Paso Real," at
which mules are swam over on the route from
San Luis to Panuco. Near the Rancho hut were a
number of high poles driven firmly into the ground

and supporting several long spars, over which some thousand yards of Tasajo were hung in long coils, to dry in the sun. We procured some yards of the beef to cook with our birds; and in the mean time an Indian woman washed our clothes in the river, and dried them on the sandy beach. In the hut I saw the hides of three light red deer, which had been killed amongst the domestic cattle, with whom they frequently associate. The number and variety of butterflies seen on this day was quite astonishing; we frequently observed several square yards entirely covered with them. They always appeared to assemble in communities of the same colour; and none which differed in tints and size ever associated together, or varied the uniformity of the bright patches, which resembled little beds of flowers.

Soon after leaving the Paso, we observed a large black mass near the top of a high bank, which the canoe-men said was a " snake with four nostrils," (*culebra de quatro narices*). I twice fired at, and fancied I had hit it, for it seemed to shake as if in pain; but it crawled into a hole before I could climb up to it, and I observed that the body was about eight feet in length and as bulky as a large orange. At noon the temperature was 84° in the shade. Gliding slowly through the stream we shot a Co-jolite, and for the third or fourth time I killed two vultures for specimens ; but, as on all former occasions, I gave up the attempt of preserving them, for it is quite impossible to convey any idea how very offensive they are. In the woods I found a tree bearing the remains of a quantity of Anona, on which the parrots, screaming around in all directions, had been for some time feasting.

Since leaving Panuco, the banks of the river had much changed in appearance, being now very steep and in many places destitute of wood. Fan Palms frequently usurped the place of other trees, and whole groves of them were at times to be seen without a trunk of any other kind amongst them. In the evening we shot some of the " Patos Reales" (royal ducks), which equal the Muscovy breed in size. They are of a glossy black, and the male has

a large white patch on the wing coverts. These birds
are remarkable for perching on lofty trees, on the
large branches of which they prefer roosting, to
settling on the ground. Heavy clouds collected at
sunset, and in the night we were drenched with
their contents, and also much tormented by the
mosquitos.

April 21.—The morning was fine, and we landed as the day broke to cook our breakfast and dry
our clothes. Here we procured one of the beautiful
birds called Calandria, whose plumage is composed
of the most richly brilliant yellow and a deep black.
It builds, or rather weaves its purse-like nest from
the extreme of a slender branch, where it swings
about with every wind that blows. We obtained
one of these curious bags, and found in its bottom
part another nest, in which were two white eggs
about the size of those of the chaffinch, and irregularly marked with spots of deep brown.

I had observed on these two last days that the
river owing to the heavy rains had increased considerably, and its current had acquired in the cen-

tre the velocity of about three miles an hour. Shoals now became more frequent, and the water was so turbid that it could not be drunk until it had stood some minutes to allow the mud to subside. At sunset we arrived at the bottom of a little ravine, above which, at about half a mile from the river, stands the small Indian village of Tanjuco. A large bevy of Indian girls were standing half naked and splashing each other in the river as we approached the shore, but soon after ran off in a hurry, laughing and talking very rapidly in the Guasteca language, which I now heard for the first time. One of the canoe-men with great difficulty procured for us admission into a hut; and we were most thankful for such shelter, although one half the room was occupied by the gear and pack-saddles of mules, and by the riding furniture of about a dozen Rancheros. Their saddles were all mounted across two long beams, with the large wooden stirrups hanging a sufficient depth below them to knock our heads whenever we moved. Four hens with broods of chickens, and three others which were sitting on their eggs, occupied various snug corners. Some half-dried beef in long odoriferous festoons dangled on one side, while on another were suspended, to dry, three raw deerskins, and the hide of a Puma or American lion. The place swarmed with mosquitos, and I went supperless to bed under my gauze curtain. My two men, who had no such luxury, were soon driven out to the open air, where they covered themselves in their blankets, and lay in company with our landlord (who was the Indian Alcalde), his wife, children, dogs and pigs, on the dusty ground.

Soon after my candle was put out, an Indian and his wife came to sleep in my room, which I now discovered was a kind of general head-quarters for travellers; and all the dogs of the establishment entering also, carried off several good things which I had hoped to have cooked for the morrow's breakfast.

April 22.—I had found the river so much swollen and so very rapid on account of the recent rains,

that our large canoe could make but little way
against it; and it was evident that a whole fort-
night would have been required to reach San Juan,
the limit to which I wished to extend my examination: I therefore determined to ride there, and
descend the stream in a small canoe. Soon after
noon I set out on mules with Manuel one of the
boatmen, and an Indian to bring back the cattle,
leaving my people to the care of the Alcalde until
my return.

On quitting Tanjuco, we entered the thicket,
where two red deer crossed our path. As far as the
turnings of our route through the closely-woven wilderness would admit of my judging, we made about
a south course, and at five came to a small cluster
of deserted huts called " Tantajou," near which
were several trees bearing a most delicately fra-
grant flower named " Flor de Rosal." In form it
resembles a large white lily, but grows in clusters
of a dozen or more on the same stalk. In half an
hour, having come south fifteen miles, we reached a
pass of the river called "Paso de Tantajo;" and I

here witnessed a most unpleasant mode of ferrying. We embarked in a very small canoe, about two feet and a half in width, with our saddles and my little luggage, having with the " canoéro " four men. Holding the halters of our four animals, we pulled them by main force into the stream,—an operation which very nearly upset the frail .hollow trunk in which we sat: but the mules soon recovered from their first alarm; and although the stream was very rapid, they struck out boldly, and towed us in a short time to the opposite bank. Paying two reals for the passage of ourselves and animals, we rode forward through an extensive Rancho, in which for three or four miles no other trees than Fan Palms were to be seen. They were thickly scattered over a large plain, where the coarse rank grass was so high that the backs of the cattle which fed amongst it were scarcely discernible.

Having ridden south-west six or seven miles, we arrived after dark at six or eight poor huts called " Tanquichi," where for some time we could not

find a living soul, or even a dog to bark at us. At
length we stumbled over a naked Indian lying on
his back on the ground, and fanning the clouds of
mosquitos from him with a cloth,—the thick smoke
of a little wood fire which was placed to windward
being insufficient to keep off these tormentors.

I may here notice a singular custom which I observed amongst the Indians and Rancheros in this
little excursion; which was, that where the mosquitos were most abundant and tormenting, they
invariably lay down stripped of their shirt; and
our canoe-men made a constant practice of this,
fanning themselves,—and I verily believe in their
sleep,—all night. Our naked friend muttered a
drowsy assent to the mules being tied to a corner
of his hut, and to our lying down wherever we
could, or following his cool example. But the women who were withinside resisted all our intreaties
that they would give us something to eat; and no
promise of money could induce them even to make
us a Tortilla.

April 23.—Having been half-devoured by the

mosquitos, by four A.M. on the 23rd we once more pursued our route through the woods. Many of the trees were of an immense size; and the Banyans in particular were quite enormous, with their several gigantic trunks supporting the monstrous branches from which they had originally descended as slender suckers. I here met with a variety of the Fan Palm seventy or eighty feet in height, but exceedingly thin in the stem, answering in appearance to the Betel-tree of the East. It is called Palma Real, and is in much request for beams and supports in houses. The people pointed out to me several wild fruits, none of which were ripe; and the ground in many places was strewed with that of the Ojite, the kernel of which the Indians boil as a substitute for maize in times of scarcity. It is a round reddish pulpy fruit, of the size of a large gooseberry, but quite smooth. In taste it is rather sweet and insipid; and it contains one seed or kernel, resembling an acorn in consistency, quite round, and no larger than a black-heart cherry. The tree which produces this fruit is very

large, and apparently an evergreen; its leaves not
unlike those of the Laurestinus, and they consti-
tute the fodder of horses and domestic cattle both
at Tampico and Panuco, where grass cannot easily
be procured.

We met several Indians with baskets of Ojite-
berries, who were going with their families to San
Vicente to hear mass and keep holiday. I observed in many places in the woods little rural
crosses constructed of the branches of trees, and
placed at the entrances or turnings of the little pathways. Here and there a large crucifix would be
seen in some conspicuous place; and all were
crowned with garlands of wild flowers by the passing Indians, who had on this day renewed several
of these beautifully simple ornaments, although
many bore chaplets of woven palm-leaves which
had long since faded.

We arrived before seven at the village of San
Vicente, at which all the neighbouring population
were assembled for church, and to be idle and
happy. I went to the hut of the Alcalde, the rally-

ing point of all the Rancheros, whose horses were tied to the posts which supported the roof. As the house was full of gossiping acquaintance, I met with no encouragement to enter it; and therefore sat down with as much patience as I could command, Manuel having resolved to hear mass. By way of diverting the intermediate time, I performed my toilette in public; shaving and washing myself, to the great edification of a numerous wondering assemblage of people who surrounded me, and who very rarely themselves performed these operations. In due time I obtained some Tortillas and Tasajo, and then went to see the church, the exterior of which I had previously taken a sketch of. It was a long mud barn, not even white-washed; a poverty of appearance which could in no way take from its sanctity as a place of worship: but it was filled with at least a hundred of the most horrid figures I ever saw, painted in gaudy colours, and varying in size from very small dolls to that of a half-grown person. One figure of our Saviour with a large brown wig was seated on a child's toy horse, exactly of the kind which our English children play with, having straight legs, and the head and curved neck cut out of a flat board. This was by no means the worst figure to be seen:—but I will not dwell on the disgusting appearance of the monsters which met my eye; such, in fact, that had a strange people visited this church, they would not have hesitated to consider the worshippers as idolaters. I can only say, that hitherto I have neither seen an original or a picture of the Mexican deities at the time of the Conquest, more abhorrent or absurd than the idols in the Romish church of San Vicente.

The Rancheros, who could not imagine what business had brought me so far out of the usual route, were very urgent that I should exhibit the goods and other articles I had brought for sale; and made offers for my gun, jacket, and in fact every thing belonging to me. On finding that I had no marketable goods, they would scarcely believe that a wish merely to see the Rio San Juan could have brought me so far.

The curiosity and suspicions of the Alcalde were so greatly excited, that he demanded my passport, observing, that " no man would ride up the banks of a river merely to descend it again by a canoe, unless he had some other object in view." I therefore showed him a Custom-house paper, which, as he was no scholar, was quite sufficient passport for me.

To escape further question I rambled over to a party of Indians, who under a large tree in the middle of the village were holding a kind of Sunday's fair. Their goods (which were spread on the grass) consisted of variously dyed locks of wool, wrappers, belts, and bad European trinkets ; and there were several parties who sold small cakes, plantains, cheeses, sugar, and other wares.

The whole of these people and the crowd which surrounded them were loitering away the day, and looking forward to a night fandango. After waiting until near noon, it was announced that the Francisca who was to perform mass was sick, and that no service was to take place. The care of this reverend man consists of about one hundred souls in San Vicente and the surrounding Ranchos; but the Pueblo or village itself does not contain above thirty or forty scattered huts. Riding seven miles we arrived at the Rancho of San Juan.

The Alcalde of San Vicente, who, I discovered, had been cross-questioning Manuel, and had come to the conclusion that I was a very important personage, overtook us on the road, and gave me his company to the Rancho. He obligingly pointed out to me several varieties of trees which I wished to see; and as our whole road lay through his property, he did this with the greater pleasure, showing me how rich he was in cedars, helping to cut a stick from the "Ule" or Indian-rubber tree, and making me acquainted with several good woods for dyeing. Some of the timber was of an immense size; but many large trunks near the road side stood dead and leafless, from having been stripped of their bark by the Indians for the purpose of tanning leather. I found the little

Rancho of San Juan a beautiful and retired spot: about two dozen huts lay scattered amongst the fine trees, and groups of cattle were straying in the long grass or sleeping in the shade of the thicket. Every thing bespoke quiet and comfort; and the simple-mannered natives appeared happy and contented in their removal from the more busy world. I obtained an entire room to myself, with a large massive table placed in a kind of holy nook, which is to be found in almost every hut in the country. In this corner may usually be seen some dozen rough and extraordinary prints, or attempts at paintings, of scriptural, or more properly monkish subjects, pasted thickly on the wall; from this is sometimes seen projecting a painted bracket for the candle, which on feast-days is burnt before the favourite saint. Crucifixes innumerable, palm branches, dust and cob-webs, generally make up the array, with not un-frequently those clusters of wasps' nests* which are so common in even the better class of dwellings.

* They vary in size from of that a walnut to half a small apple,

The mistress of my room (who lived in a hut opposite) soon discovered that I was hungry, and sent me some palmito root very nicely prepared for my dinner.—This is the tender heart of the young fan palm before it has acquired any stem, and is about the size and thickness of a man's leg. It is very excellent when cooked, and even while raw has an agreeable flavour somewhat sweeter than that of a white cabbage, which it much resembles in crispness and consistency.

I found in the hut the skins of two leopards which had been killed in the Rancho. I also fished out an imperfect piece of sculpture, bearing some resemblance to the lion-figure-head of a ship, and heard of several more at an ancient city some few leagues distant called " Quaï-a-lam."

I had never until this day seen that troublesome insect a " Nigua," or Jigger, and then saw

___

and are built entirely of clay, in the manner of the swallow's nest. Each, according to its size, contains a number of little cells for the larvæ of the wasp, which are plastered up until they have strength enough to burst their passage out.

quite enough of them in the toes and feet of a
poor little pot-bellied Indian orphan belonging to
the house, and comptroller of the hens and chickens. He was sitting with one leg under him like
a China Joss, and the other stretched forward into
the lap of a girl, who with a blunt needle was extracting the nests of these insects, which are inclosed in tough white bags sometimes equalling a
rape-seed in size. During this operation the boy
never winced, so that from his apparent apathy I
concluded him to be an idiot; but the poor thing
was weakly and ill, and every part of him but his
prodigious belly had been wasted away by the
fever which carried off his parents. He appeared
to take a great fancy to me,—I verily believe from
my having made so many compassionate faces for
him when I saw the girl tearing his toes; so I
gave him a medio, which was the first money he
had ever possessed, and he in gratitude sat and
stared at me all the evening. I was much amused
by the way in which I obtained information as to
the age of this " Cristiano," who himself knew

nothing of the matter: but a woman hearing the inquiry, told him in the language of a farrier to show his teeth,—and pronounced him to be six years old.

I believe that every woman and the greater part of the men of the Rancho came at different times in the evening to see my watch and writing-case, neither of which curiosities had ever before been exhibited in San Juan. The watch was a machine of which all had heard; but their astonishment on hearing it tick and seeing its wheels in motion, was really as great as I ever saw displayed by either Negroes or Esquimaux; yet these people were almost all white, and the descendants of Spaniards.

A venerable old Ranchero whose opinions seemed to carry great weight, remarked, that "it was a folly (*touteria*) to give a number of dollars for a thing just to know how many hours it was from morning or night;—that to know when to eat and drink, when to get up or lie down to rest, was quite sufficient:" a remark which with these primitive people met with very general approval. Offers were made to purchase every thing belonging to me;—a mano of paper, buttons, any article whatsoever; my visitors being pursuaded that I could come for no other purpose than to trade.

The poor people at San Juan have no means of procuring foreign clothing, arms, or utensils, but by sending down the river to Tampico; and they are in consequence obliged to pay so exorbitantly, that the presents I made them of paper, a pencil, and other trifles, gave them a high opinion of my generosity.

It is remarkable that there are no mosquitos at this place, although above and below it, and in fact in almost the whole vicinity of the river, they are exceedingly numerous and tormenting. I know not to what cause their absence may be attributed, as the temperature, the relative situation and the soil, are precisely the same as at those places most infested.

April 24.—On waking in the morning I found

my landlady's daughter standing by my side with sweet bread, and coffee boiled with milk and the pure expressed juice of the sugar-cane,—a preparation I have never seen excelled as a breakfast. I then joined a party of Rancheros who had assembled to kill a cow and cut her flesh into tasajo,—an operation which they performed with extraordinary skill and dispatch, separating the sinews from the flesh with anatomical precision. Two men devoted themselves to cutting the meat into long strings or ropes, which they threw to another who rubbed them well with salt: after this, no other process remained but to hang the beef in festoons over long poles to dry in the sun.

I breakfasted with the Rancheros, when their work was done, on dry meat with Chili sauce and piping hot tortillas served up in rapid succession. Our second course was a dish of cow's blood stewed with sweet herbs: and having prefaced our meal by a glass of white brandy distilled in the Rancho, we all ate heartily. When the meal was over, a little boy who had waited on us kneeled down facing the picture corner of the room, and recited a prayer and thanks to the Virgin *. My companions devoutly crossed themselves, and then proceeded to examine my curiosities. They were wild good-natured fellows, with unshaven chins and greasy leather clothes; and we formed a most merry group, I being delighted to have it in my power to amuse them.

After very great difficulty I procured a canoe; but as it had not been in the water for two years, the bottom was split and in sad condition. With petroleum and slips of wood, however, we put all to rights, and at 4 P. M. started on our return

* This simple and sincere act of devotion is very general amongst the country people, who hear the grace recited by the youngest person in company, or by a man-servant if they have one. All then cross themselves, and bowing right and left salute their neighbours with " Bon provecho !" (Much good may it do you !) I had a servant who never omitted this ceremony at my frugal meals as I travelled through the country; and although his prayer was in Latin, rapidly uttered, and neither understood by himself or his master, it was an act of unaffected piety which I respected too much to interfere with, even though he frequently invoked the protection of Our Lady of Guadalupe, and sometimes that of Señor San Francisco.

for Tanjuco, Manuel and a young Indian taking the paddles.

I left my old landlady and the pretty Rancho with regret, having met with great kindness and attention from the one, and found rest from the mosquitos in the other. I gave my hostess at parting a white neckcloth and a pair of scissars; and carried off many good wishes, with a large supply of tortillas and stewed meat and a bottle of white brandy; for all which, with the subsistence of Manuel and myself for two days, I only paid about six shillings.

It may be requisite here to explain, that much of the details of courses and situations of Ranchos and particular points is omitted in this Journal, in which my object is merely to relate my personal adventures.

Near San Juan the river is narrow, extremely turbid, and full of impetuous eddies and rapids, in which many sunken trees just show their situation by the water breaking over them. The people managed the canoe very well in this troubled navigation; and at night-fall we hung the boat to a grounded tree in the middle of the stream, and slept until daylight very comfortably, the canoe exactly fitting us in width and length.

April 25.—While shooting an impetuous and turbulent rapid, we whirled past the nose of a large cayman which lay basking on the sunny bank. It was the first which had appeared for many days; and as we were swept unresistingly with the stream amongst the shoals and stumps of timber, I could not but feel how very unpleasant it would be to upset the canoe and to find oneself in the same element with the creature we had just passed.

My patience became exhausted by sitting for many hours in one posture to maintain my equilibrium, under a temperature of 109°; but I was at last somewhat consoled by finding an alligator's nest with thirty-nine eggs. It is the custom of the caymans to select some sunny sandy beach, in which they bury their eggs, piling a large heap of sand above them. They then

leave their offspring to be hatched by the heat of
the sun, although, as the Indians informed me, they
keep " a register in their head," and return at the
expiration of thirty days, when their newly produced
little ones are ready to be taken on the mother's
back and receive their first lessons in swimming.
The idea that the alligator devours her young if
she can catch them, is denied by the Indians, who
on the contrary declare her to be very kind to
them. I should like to have seen in what way the
maternal solicitude of one of these horrid creatures is shown; for a nursing alligatress must be a
great curiosity. The eggs are about the size of
those of our domestic ducks, but bearing a highly
enamelled surface. At each end they are translu-
cent, but an opake white band encircles the middle, which appears to have a dividing membrane
across it. The yolk also resembles that of a duck's
egg, but has a slight flavour of musk, and the
white is nearly of the consistence of jelly. On the
border of the river I shot a small eagle, of that
species which, according to the Indians, preys

upon serpents. In landing for my bird, I crossed the recent track of a leopard which had been drinking in the stream, and in a few minutes killed two fine turkeys. We this day, at various times, had passed a great number of Indians, who were bathing in the river by whole families at a time, which appears to be their morning and evening custom; and all those who live near the stream are very clean both in their persons and clothes. Boats laden with articles for the Tampico market continually enlivened the scene; and it was highly amusing to observe the politeness of the passing Indians, who used the " Don" and pulled off their hats very ceremoniously to each other on every occasion, paying a variety of rapidly uttered compliments at the same time. The señoras and señoritas who sat washing themselves or their clothes in the river, received the most marked respect. Many a brown flat-visaged man, with a quarter of a pair of breeches and a straw hat, was hailed as " Don;" while inquiries were made after the health of the señora and the young ladies, who in some instances answered for themselves as they were disporting in the water near the banks, and just showing their shining brown shoulders and immensely long jetty hair, while they swam in those places unfrequented by the alligators.

At sunset we passed the entrance of the small river Tempoál, which flows into the San Juan from the S. S. E., and whose perfectly transparent water formed a curious contrast with the muddy stream in which we were paddling. We landed for the night on a little sandy point, where myriads of mosquitos hailed our arrival with the most rapturous humming; but we kept them at bay for a time, by making a large fire to scare the lions and tigers *, of which the canoe-men had, I think, very groundless fears. We here cooked our supper and broiled some ears of maize, on which we made an excellent meal; and then resigned our persons to the hungry insects, which in their turn fared

---

* The puma, and a large species of leopard are so distinguished by the native Mexicans, who have no idea of the existence of more ferocious animals of these names.

most sumptuously. It may not be amiss to state here, that by determining never to rub a mosquito bite,—inflammation, swelling, and a great stock of extra pain may be avoided. The people, in consequence of having scratched themselves, returned covered with sores, while I on the contrary, having refrained from this indulgence, retained scarcely a mark *.

April 26.—We pursued our journey at daybreak, and I soon forgot my sufferings in the pleasure of shooting four turkey-cocks. I saw also a great number of rabbits, cojolites, doves and other birds, as we glided down the river; but I was too rich in turkeys to pay any attention to them, and was anxious to reach Tanjuco before night.

* All my experience of these tormentors proved to me that they have not degenerated since the time of the Conquest; when Bernal Diaz most pathetically describes them, and almost ranks them with the Indians, saying that in addition to the natives, " habia muchos murcielagos, é chinches y mosquitos, é todo les daba guerra " (there were also many bats and bugs and mosquitos, all of which made war upon the soldiers).

At noon we passed some high cliffs, which being of a reddish earth are called Tierra Colorada. They are but scantily wooded, and are said to be the favourite resort of pumas, leopards, jackals, and the jabali or Mexican hog.

At twelve miles above Tanjuco and fifty-nine above Panuco we passed the junction of the river Tamoin * with the San Juan, where their combined streams receive the same name, and flow to the sea as the Rio Panuco. After stopping for a few minutes at the Rancho of Bichinchijol to procure some sour curds and sugar from a sweetheart of Manuel's, we made the best of our way to Tanjuco which we reached at 4 P. M.

In this little excursion I had fully verified the assertions of my boatmen as to the difficulty of ascending the river in our large canoe, the rapids being so strong that even our small one had passed some of them with considerable risk. After the recent rains the current had become very im-

* Here about sixty yards in width, the united stream is about one hundred and fifty.

petuous, and the quantity of sunken trees which lie in the narrow parts of the river renders the navigation difficult and dangerous.

As far as my hasty visit would permit me to judge, the only point to which laden boats of any size could ascend with safety is Tanquichi, about thirty miles below San Juan; whence, should any commerce turn that way, mules could always be procured to any other part of the country, the paths through the forests and thickets being said to be very good.

From San Juan to Tanjuco, by the most accurate compass bearings and estimated distances, is seventy-eight miles, in which distance the river takes no less than one hundred and six very abrupt turnings. The general width of the Panuco is from half a mile to one hundred yards; that of the San Juan is in many places much narrower. At forty-six miles above Tanjuco and ninety-three from Panuco, the Rio Tempoál joins the San Juan.

I found that my two Englishmen had been very

kindly treated by the Indians, with whom they had made themselves on excellent terms, although neither party understood one word that the other said. Owing to the arrival of a kind of pedlar with dresses and various articles for sale, there had been great doings and fandangos in the village every night, the strangers always being invited to join the dancers under a large shed, which acted as a cow- and sheep-house by day, and was transformed into a ball-room at night. The old Alcalde made a point of being very drunk and important on all these occasions, and no dancing took place until he thought proper to fire his musket, which was the anxiously awaited signal for the merriment to commence.

Tanjuco does not contain above twenty very miserable huts, well stocked with naked children and pigs; and as it has not a single white inhabitant, and very few speak or understand a word of Spanish, it may be considered as a good specimen of what the Indian towns were at the time of the Conquest. The huts are still built of stakes and bamboos, and thatched with palm leaves. No traces of Christianity are to be seen, the nearest church being nine leagues distant, at Panuco; and the perfect mildness and simplicity of the natives not a little favours the resemblance to the settlements of which the early writers have given us a description.

April 27.—We left the village early in the morning, having laid in a large store of watermelons; a luxury of which the value can only be estimated under a temperature of 90° in the shade, or rather under the reflected heat of our hide-awning.

Passing a steep cliff, fifteen miles from any village, we observed a quantity of earthenware to be mingled with its clay; and one of the canoe-men discovered a small crockery-ware leg protruding itself. After securing this, at a few yards further we picked out the body of a curious little female figure, also in terra cotta.

Having now no further bearings to take, we slept and drove with the current; and soon after noon on the 28th arrived at Panuco, where two American schooners and a Campeachy brig were lying at anchor; the latter discharging a cargo of salt, for which there is great demand by the people who prepare Tasajo. Waiting until the cool of the evening, we again dropped down the river.

April 29.—At breakfast-time on the following morning we landed at Tamanti to purchase some milk and half a stone goddess, of which I had heard at Panuco, and which was a heavy load for the four men who carried her to the canoe. She now has the honour of associating with some Egyptian idols in the Ashmolean Museum at Oxford. We landed a second time for a supply of the clustering limes which hung invitingly over the stream; and then paddling all night, cheered by the prospect of a bed and fewer mosquitos, we arrived early on the morning of the 30th at Las Tamaulipas, where the gentleman whose return from Real del Monte I had been directed to await, had arrived two days before me. On returning to my old quarters, I found that I had sustained a sad

loss in upwards of one hundred carefully prepared
specimens of birds, which my servant and I had
at various leisure moments procured with great
exposure to the burning climate. The small black
larvæ of some prolific moth had got amongst them
in prodigious numbers, and even the scaly skin
of the alligator had not escaped their destructive
ravages.

Unfortunately for my repose after my expedition, I found that the Commandant with a large
party of his officers had engaged themselves to
dine on board the Company's brig. Fatigue and
a head-ache pleaded my excuse for not joining
them; but in the evening they became so very
drunk and clamorous for me, that I was obliged
to comply, and was almost immediately forced to
accompany them in their boat to the other town.
My companions now began splashing each other;
next they proceeded to bale hats-full of water in
all directions; and finally, when we reached the
spot where the caymans most abound, they amused
themselves by throwing somersets into the river,

an operation which nearly drowned a poor old captain lineally descended from one of the Conquistadores. I now thought proper to interfere, and made the crew land at a place half a league from the town; a distance I shall well remember, as it fell to my lot to carry old Garibai to his house, an effort from which I did not recover for many days.

May 1.—His Majesty's ship Tweed having anchored off the Bar, I went on board, and paid a longer visit to Captain Hunn than I had anticipated; for a summer norther set in with a very heavy sea, and I was unable to land until the 3rd, when the Tweed sailed again.

May 4.—I accompanied a large party to Tampico on the afternoon of the 4th, in order to be present at the ceremonies of the Feast of Ascension. For some previous days the natives from every part of the surrounding country had been flocking in to assist at this annual and most important festival. The towns of Pueblo Viejo and

Las Tamaulipas were almost forsaken, and we found the little village of Tampico crowded with many thousand people in their gala dresses. The Plaza was filled by booths containing clothing, trinkets, toys, cakes and other articles, just in the manner of our country fairs, but there were also several gaming-tables under canopies. The Commandant with two companies of his troops in their best uniforms were also there, and the whole scene was really very lively and interesting.

At three o'clock the two bells of the Parroquia tolled rapidly; and the crowd gathering round the church, we entered with them to where the old priest stood by the tawdry crucifix of which I have already spoken, in order that he might touch the leg of the image with ribbons and crosses which were brought to be thus sanctified as reliquias in return for an offering proportioned to the zeal of the giver. Some therefore subscribed very largely; and a heavy dollar thrown from the hand of a poor Indian, who perhaps did not possess an-

other in the world, would frequently jingle amongst
the lighter contributions of less pious Christians.
Many obtained " Indulgencias" by kissing the
newly painted thigh of the statue ; and it was not
until the subscription ceased, that the figure was
lifted and carried from the church on the shoulders
of voluntary bearers habited in white robes.

At this awful moment a ship is annually seen by
the eyes of the Faithful, far away in the offing,
bringing, to bless the solemnities, that Christ whose
miraculous image is moving in procession ! Hundreds now leapt upon the walls near the church,
with open mouths and straining eyes, to look out
upon the distant horizon; but none of those
whom I questioned would venture to affirm that
they actually saw the ship. The image was paraded round the whole town, preceded by two
men, who gratuitously read aloud before it. The
head of one of these people was too remarkable
not to be noticed, and I observed that he chaunted
with much greater vigour than his companion,
and apparently in a different form of words.

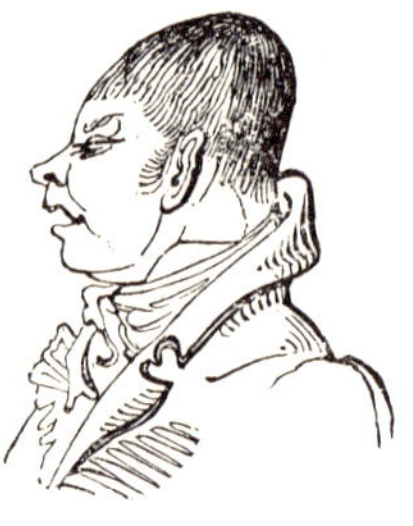

Further in advance marched a most singularly
assorted band of musicians.   There was a harper,
with fifers, abundant Indian rattles, a woman and
several men playing on fiddles, each performing
his own tune.   When all was over, the people dispersed to dance and gamble; and we rode back
again to the old town.   Hundreds of weary women were on the same road, carrying bundles containing their holiday finery; and many who had
worn white satin shoes, silk stockings, and gowns
of the finest French materials, with very rich veils
at the Fiesta, were now returning home bare-footed
and merely in a shirt and petticoat.   I could not
but remark throughout the day the very striking
difference between this and our English festivals,

where all is bustle, noise and activity, though mingled, it must be confessed, with a few broken heads
and black eyes.   Here, an air of quiet lazy happiness seemed to pervade all classes.   The men and
women sauntered slowly through the village, where
many lay sleeping on mats in the shade, while
others were dispersed in listless smoking groups
under the little awnings of the booths.

On the 7th a vessel arrived with people for the
mines of Catorce, and I received by her several
letters which much affected my future plans in
Mexico.   A young gentleman who came in this
vessel was in a dangerous stage of yellow fever,
which he had contracted by sleeping only one night
on shore at Vera Cruz. He died on the 10th. The
Spanish surgeon who was called in, shrugged his
shoulders, and plainly told the patient, even at the
first visit, that his case was hopeless; nor would
he prescribe any medicines for him, considering
them of no use.   Mr. Price, a medical gentleman
attached to the same Company with myself, also
visited the unfortunate invalid, though it was then

too late, and attributed his death in a great measure to the want of proper and timely assistance.

May 12.—The brig of the Real del Monte Company crossed the Bar; but by the ignorance and stubbornness of the pilot she tailed on the northern shoal, and carried away some of her false keel. At this time the depth over the bar was thirteen feet. I was fortunate enough on this day, in crossing to Pueblo Viejo, to see a Manati, or Sea-cow*, which, having been left by the tide in shoal water, had been killed and drawn on shore. Its body to the commencement of the tail was eight feet and a half in length, and the tail itself, shaped like a broad paddle, two feet and a half more; the latter was quite flat, placed horizontally as in all the cetaceous animals, and three feet across. The carcase was about equal to that of an ordinary seahorse or walrus, and might weigh about ten hundred weight; and in form, particularly about the head, it was very similar to that animal. The fins or arms, of which it had only two, were shaped

* Called also Lamantin: *Manatus Americanus,* Cuv.

much like those of the whale. There were but few hairs on the skin, which when deprived of the cuticle greatly resembled in exterior appearance prepared hog's hide. While wet and recent it was about an inch in thickness, and I brought away some thongs, of which I made strong whips.

The flesh, in colour, was like that of a raw turkey; the fat not rancid, but on the contrary very sweet; and a steak which I cooked tasted so like well-fed young pork, that the difference could scarcely have been detected.

In the afternoon our brig sailed for England; and I now after a series of most provoking detentions prepared to start for the interior.

May 14.—All my accounts were now settled; and as I was prepared to leave Tamaulipas on the morrow, I threw together a few notes respecting this part of the country, which I had omitted to insert in the course of my diary.

The maps and general opinion would lead strangers to imagine that there was but one Tampico, when in fact there are three towns on the Rio Panuco which bear that name, either singly or combined with some other appellation. The first and most ancient, as its name implies, is Pueblo Viejo de Tampico, situated on the border of the large shallow lake which lies to the southward of the river. This was originally a cluster of the huts of fishermen, who supported themselves by catching, in the months of May and June, vast quantities of a large species of shrimp, which being dried became a considerable article of exportation. These shell-fish are still as abundantly caught as ever, but for very many years the town has been important in a mercantile point of view. It is recorded that Admiral Drake once visited this place, and carried off all the wealth of the inhabitants *, which induced them to found the village now distinguished by the name of Tampico, and situated on a rising ground amidst the thickets about seven miles to the southward. This latter place never arrived

* The memory of this visit has been preserved in a singular way, although I have but little confidence in the authenticity of the story. The English sailors are said to have introduced to the natives whom they plundered, the method of making grog, which name having I suppose been too difficult to remember, has been supplanted by that of *Drak*, in memory of our English admiral.

at any degree of importance, and Pueblo Viejo
has again become the chief town. It has at present a population of about 4000 souls, the greater
part of whom owe their subsistence to foreign trade.
The houses are of a very inferior description, but
the shops are many of them well stocked with European commodities. One of these depôts bears
the name of "The Deity," and others are no less
blasphemously distinguished. There is a large windowless mud church here, which I rarely saw opened. The two inns are miserable, dirty and comfortless; but from subsequent experience I can declare that they are, with the exception of one recently opened at Vera Cruz, the very best in the
whole Republic of Mexico.

The market is tolerably well provided with meat,
fruit and vegetables, the latter of which are brought
from a considerable distance up the river in canoes.
Its supply of fish is but little attended to, although
the Rio Panuco abounds in them. Turtle, which
are large and numerous, are not esteemed; and the
periodical fishery for shrimps is, as I have said, the

only thing now attended to. These are taken by means of wears and large hoop nets, and during their season the little passages leading from the river to the lake of Tampico are almost stopped up by the shrimpers. Game is abundantly supplied to the market, and consists of ducks, wild turkeys, the large crested pheasant, chachalacas, cojolites, and other kinds; but venison is not often procured.

The vicinity of the shallow lake is considered as the principal cause of the agues and fevers which are so prevalent at this place, the water being at certain seasons so low as to leave extensive muddy shoals and oyster banks exposed to the burning rays of the sun. On landing at Pueblo Viejo a stranger becomes immediately impressed with the idea of being in an unhealthy place, by seeing the house-tops covered with flocks of the black disgust-ing-looking carrion vultures, or by finding them battening on their offensive food in his immediate path, from whence they are not easily dislodged.

The immediate borders of the lake, and the site of Pueblo Viejo are remarkable as being formed

of a stratum of decomposing oyster-shells, which
is of considerable depth, although the impossibility
of digging below the level of the waters of the lake
in sinking wells, precluded my seeing an extent of
more than twelve feet.

The next town in point of importance, and which
will soon eclipse the other two, is the Pueblo Nuevo de las Tamaulipas, situated about three miles
to the northward of the old town, and on the bank
of the river. This, as I have already said, is of very
recent erection, and increasing rapidly, being built
on a neat model, and inviting by its commercial
advantages the removal of the natives from Altamira and other parts. There can be little doubt
of this having been the situation of the populous
town which was discovered by Juan de Grijalva
in 1518, when the warlike natives attacked his
ships with so much courage, and were beaten off
with considerable difficulty*. The historians of

* In consequence of the assailants coming to battle in sixteen large canoes, the river at that period obtained the name
of " Rio de Canoas."

VOL. I.

the Conquest of Mexico all agree in saying that the wars in the country near the Panuco were carried on by Cortez' soldiers with the greatest fury; and it is not improbable that the town at Tamaulipas was the first destroyed by the invaders. Remains of utensils, statues, weapons, and even skeletons, have frequently been discovered in digging the foundations of the recently erected buildings; and some singularly shaped figures in terra cotta, which I now have in my possession, bear the closest resemblance to others I have met with in more distant provinces.

In the Pueblo Viejo and Pueblo Nuevo the want of potable water is very severely felt; and the inhabitants receive their chief supply from the Tamesi, a small stream which branches from the lake of Altamira, and is about three leagues up the river Panuco. Large canoes are constantly employed on this business, and it is customary for families to send relays of small casks according to their wants.

Another great inconvenience is the extreme

scarcity and dearness of fodder for horses, which are fed on the leaves and young branches of a tree named Ojite, growing in considerable abundance at some distance in the woods.   Notwithstanding the vicinity of immense herds of cattle, milk is rarely to be procured at Tampico: fresh butter is never seen, in consequence of the heat of the climate; and the only cheese brought to market is an inferior kind of preparation of curds.

The Tampico towns have very materially risen in importance since the establishment of Mexican independence; and a brisk trade is constantly carried on with the United States, whose small vessels have great facilities in passing over a bar, which offers a sad impediment to our more weighty merchantmen.

CHAPTER II.

Journey from Las Tamaulipas to San Luis Potosi—Santa Barbara —Tortillas — Holy Picture of the Virgin — Ancient Building—Tula—Funeral of a Child—Peotillas—San Luis Potosi—Route from San Luis to Zacatecas—Vino Mescal—Salt Marshes —Arrival at Veta Grunde of Zacatecas.

AT seven on the morning of the 15th of May I left Las Tamaulipas for San Luis Potosi, my party consisting of my servant and another Englishman, a native attendant named Flores, with four Arrieros or muleteers, and fifteen mules.

We set out on the road to Altamira, already described as being through a closely interwoven forest, and at one we reached the little town. The muleteers deposited their cargoes on an open space near the Campo Santo, while we placed our saddles and persons in a roofless hut about half a mile from the town, in one corner of which a sick Indian, attended by his wife, was moaning with fever and ague, while muleteers, dogs and vermin surrounded us. These were inconveniences to which every traveller in Mexico must endeavour to reconcile himself; since, with the exception of the high road from Vera Cruz to the capital, he has, particularly in the northern states, to traverse a country which may literally be called one continued wilderness.

May 16.—Leaving Altamira at an early hour, the road was very good, with broad beaten paths amongst a few scattered thickets. We had not ridden far when I found two wild hogs (Jabali) lying by the road side with their heads cleft in two, as if by the strokes of a hatchet. None of their flesh had been taken away *, although they were

* It may perhaps be superfluous to remind the reader, that these animals are mentioned in old books of travels and natural history, and still by the natives of the country, as having their navel on the back. There is a gland on this part whence issues a secretion, which gives, if the part be not cut out immediately after death, a most unpleasant flavour to the flesh ; and it is this peculiarity which has given rise to the idea of the Peccari, or Mexican Hog, having its navel on the back.

fat and in good condition; but a large flock of vultures were enjoying a most savoury repast upon it.

After a ride of three leagues we arrived at a solitary hut, " La Potranca," where we obtained some curds and whey; in asking for which I made an odd mistake, by inquiring of a very old woman if she had any " Suégro" (father-in-law), instead of " Suéro" (whey). Leaving this place, Flores pointed out to me in a small thicket, the grave of a young Frenchman, who about six months before had shot himself while in a state of delirium caused by fever.

The country now became rather uneven; the scenery was more interesting, and enlivened by a great number of light-coloured hares, which were running fearlessly near us.

Riding seven leagues we reached an Estero,—a long narrow pool of turbid water,—of which the animals drank eagerly, the thermometer being 110° in the sun. In three leagues more we arrived at a lone hut in the middle of a plain, dignified by the name of Guarita*. I decided on resting here for the night, as my mules were far behind, and the baggage had to pass an examination by a custom-house officer and two soldiers. At this place as well as at almost every other rest we made, we all slept in the open air, which, whatever may be reported of its insalubrity, is at this season far more refreshing and agreeable than lying under cover. Deer are so very plentiful on this plain as to be frequently seen in large herds, going to drink at sunset from a pool of water near the hut; the inmates of which said it was too much trouble to shoot them, although by their own confession they rarely enjoyed the luxury of meat with their Tortillas.

May 17.—Soon after seven we set out, and in the cool of the morning saw several wild turkeys, hares, and two varieties of quails. I killed with my whip one of the beautiful coral snakes reputed to be so venomous. This reptile was alternately bar-

* Custom-house station.

red with deep red, and black bands of nearly equal
width, the latter being edged with narrow yellow
borders.   It was about three feet in length, and
the poison fangs, which I extracted, were very
large.   We soon afterwards started a very long
snake, called "Alicante," which is said to be perfectly innoxious; but after a chase of half an hour
it escaped us.

In the forenoon the Bernal of Orcasitas (or
Horcasistas) suddenly appeared before us as we
ascended a gentle eminence.   It was still very
distant, but the outline of its remarkable summit
appeared distinctly above some light fleecy clouds
which floated round its base.

The intermediate plains covered alternately with
long yellow parched-up grass and low thickets,
rendered this first view of the Bernal very striking.   At two leagues from the Guarita we reached
"La Primera Puerta del Chocoyo," a name applied to two hovels, as well as to the gate of a large
inclosed Rancho, communicating at about fifteen
miles on the right with the Hacienda del Cojo.

Two leagues further westward we reached a narrow pool of water, called "Charco Largo," and then passed "La Segunda Puerta" with two other huts; after which, in three leagues, we reached the little Rancho village of Charyssa,—a miserable assemblage of huts on a most disagreeably dusty place. The heat in the shade 93°.

Here the Arrieros persuaded me to stop after our burning dusty ride; and I unfortunately selected for a resting-place a space between two little mud cabins, in each of which were sick people,—one old and two young women, who moaned and talked in a whining half-crying tone all night; a constant custom with the Mexicans, whether seriously ill, or only slightly indisposed.

The old lady, to whom I gave a couple of pills in order to remove her head-ache, very deliberately poked one up each nostril, as being nearer the seat of pain; but a little explanation procured their extraction, and insured the transfer to her mouth.

One of my servants too had suffered so much

from the heat of the day, that in the evening a severe attack of fever came on; but a copious bleeding, which astonished the natives, and some appropriate medicine, arrested its progress. The
arrival of a foreigner was soon rumoured about
the huts; and in consequence many intended
purchasers of paper, knives, trinkets, wine and
clothing, crowded round me, and would scarcely
believe that I had not a pedlar's pack to open for
their inspection.

May 18.—At a mile from the Rancho we crossed a small muddy stream named " Rio de Charyssa." It was at this season scarcely deep enough
to cover its stony bed, and flowing from the northward ultimately joins the Tamoin, which enters
the Rio Panuco. The morning was cool and over-
cast, and the majestic Bernal, whose base we had
now approached, reared its rugged head above the
zone of dark gray clouds which floated in dense
masses around it. The only strong light it received was from an occasional ray of the rising
sun, which gleamed for a time on the edges of the

clouds, and gave an air of indescribable grandeur to the scene. Our road led us to the left of the Bernal, and at about seven miles from its centre; so that I was enabled to obtain a good view of it, which may be better understood by the annexed outline sketch.

The base stretches about four miles, and rises very gradually to the centre, from whence an immense cluster of naked rocks rise to about two thousand feet above the level of the plain; while not another rock or mountain is to be found within at least thirty miles of the Bernal. The morning soon became sultry, the temperature in the shade being 94°, but it was one of the most animated I had enjoyed. Immense herds of cattle which had been driven in to be milked, were returning to their pasturage on the plains. A considerable number of these were very large animals, and of a light dun colour, evidently of a peculiar breed, as I never saw any others like them in the country. Many hundred mules came in from their night's graze, to receive their cargoes, which lay dispersed in groups upon the plain, attended by crowds of Arrieros, who were carrying a large convoy to the interior; and a herd of half-wild mares and foals swept past us at a gallop on their way from being driven to water at the river. Flocks of sheep, goats, and asses formed the more quiet part of the picture; while several Vaqueros, or herdsmen, were careering about and lasoing such animals as had strayed.

The country since leaving Tampico had been uniformly level, and the roads, with the exception of that to Altamira, excellent. There were some few plains quite destitute of trees, and thickly covered with the rank coarse grass, which at this season was quite parched and as dry as straw; but generally the country was clothed with a close thicket of low stunted shrubs and mimosas, and no trees of any girth of timber were to be seen. This may be accounted for by the excessive heat and drought of the Tierra Caliente during more than half the year. I remarked, however, that an inferior kind of the Tuna (a *Cactus*), which is a very juicy plant, flourished more abundantly than any other; and the dry tracts of the low lands of Mexico, as well as the arid plains of the great African desert, are remarkable for bearing a few species of the more succulent vegetables. Some few varieties of Palms, Yuccas, and dwarf Aloes, were also to be seen scattered in the dells: and the Arrieros pointed out to me a small plant somewhat resembling a young birch, and named " Palo de Leche" (milk-wood, or stick), from its property of yielding readily, if slightly struck by a whip, a very opake white juice; believed, I know not how truly, to possess a poisonous quality. At all events, it forms an interesting subject to the muleteers, who whiled away the fatigues of our sunny ride by recounting to me a variety of horrid stories of the uses to which this milk had been applied, either to satisfy jealousy or revenge, although all my informants confessed their inability to vouch for the truth of these recitals. After a ride of about seven leagues, we reached a small stream named San Juan, which was at this time very low, and had formed itself into many shady little pools, at which the mules drank; and I remained behind to enjoy the luxury of bathing. I would advise any traveller who might follow me over this road, on coming to San Juan, to look out for a round pool of water at about one hundred yards to the left of the cross-ing place, shaded by fine evergreens which dip their branches in the stream. There is a large flat stone on which he may place his clothes; and if he has the good fortune to possess a cake of soap, he may make quite another man of himself, after all his dusty scorching journeyings from Tamaulipas. Let him however remember, on being left be-hind, that two roads lead from the water, and that he must take that to the left; by going to the right I lost my way, and gave a great deal of trou-ble before I was found again. In two leagues from hence we reached the Villa* of Horcasistas, consisting of about one hundred mud huts and houses, and having a large barn-shaped church, in front of which two bells were suspended on a scaffold. Here we obtained some dinner, and purchased also provisions for supper, as we were not to see another house until the following day. Riding six miles, we came to a small pool of water called "Arroyo de la Laja," and in four miles further, through some delightfully shaded paths, like our little English lanes, reached a small bare space amongst the thickets†, where we pitched for the night. Close on our right was a deep gully, half choked by luxuriant plants and abounding with very good water.

Our road since leaving Horcasistas had become rather uneven, over hills and dales, for the greater part covered with trees; and our general course from Altamira had hitherto been about W.N.W.

Making our beds under a clear blue sky, we all

* Here implying a town, in contradistinction to a village.
† Called from a neighbouring brook, Arroyo del Sargento.

turned in very comfortably; but soon after mid-night a rain-cloud burst over us, and a thorough soaking was but the operation of a minute. With all the misery of a forced ablution of this kind, the sufferer can scarcely refrain from laughing at himself and companions, as he stands dripping in his shirt, and watching the fruitless searches made for sheltered nooks amongst the baggage, which perhaps can only be reached by paddling through the thin mud, while it is splashed in all directions by the tropical torrents.

May 19.—Two mules were lost for some time in the woods, which enabled us to dry our clothes before they were found; and in the mean time I enjoyed a bathe in the Arroyo (or brook), which gives the name to the place.

After riding about five miles, I remembered having left my thermometer by the water side, where it stood at 92°; and riding back to seek for it, met two strange Arrieros, who having found a valuable sword which had dropped from beneath my housings at our outset, were endeavouring to come up with our party to deliver it to the proper owner. I gave them all the money I had about me, lamenting that I was so far removed from my baggage as to be unable to reward so much honesty as it deserved.

In eight leagues from the Arroyo we arrived on the steep picturesque bank of the Rio Limon, a clear beautiful stream about two hundred yards wide, at this turn running to the eastward over several small falls which forbade its navigation even by canoes. Riding two miles further through the woods we again came to the river, and crossed it at a ford about three feet deep, between two small rapids, above and below which alligators sometimes resort. This river joins the Tamoin, and in the rainy seasons is said to be very impetuous, and dangerous to cross, scarcely a year passing without some people being drowned at the fording-place. We ascended the bank to a cluster of Rancho huts, amongst which we piled our baggage for the night, bought a kid, and cooked our supper. We had travelled about thirty-six miles over a good but uninteresting road, through a close wilderness, chiefly of mimosas, yielding abundance of gum, which would well repay any persons who would be at the trouble of collecting it. The Mesquiti (an acacia) is the tree that yields most of this substance, which I believe to be the gum Arabic. The long narrow pods of the mesquiti have when ripe a most agreeable saccharine flavour, somewhat resembling that of the Locust-bean, so common in the Mediterranean. The wood of the tree is very close and hard, and the charcoal made from it is considered superior to any other. The temperature in the shade was 98°, in the sun 114°; and it continued throughout the night at 80°.

May 20.—Setting out at 8 o'clock, and riding through a forest of fan palms, tall bamboos, and fine timber-trees, we came to a transparent little stream, near which three deer of reddish colour and the size of our fallow-deer were feeding; and in five miles more stopped to water our cattle in another rivulet, gurgling over flat ledges of clay slate, on which there was not the slightest deposit of mud or weeds, and closely overshadowed by fine evergreens. I lay down on one of these ledges

under a shady tree, and let the waters of " La Ri-
chuela" flow over me for half an hour. Riding
six miles, we ascended the steep rocky " Cerro de
las Cucharas" by a bad and difficult path, under
a heat of 94° in the shade; and again descending
to a plain covered with fan palms and acacias, in
about twelve miles reached a few huts called "Chamal," on the beginning of the first ascent to the
Table-land*. We had the precaution to put up
our tent here, and at midnight heard the waters
from a thunder-cloud rush like a river past us.

May 21.—We set out very early in the morning to ascend the " Cuesta de Chamal" by a rocky
and fatiguing path, rising, I should imagine, about
a thousand feet above the village. In four weary
miles we passed this " mule-fatiguing" place and
crossed a smaller Cerro, whence we descended to
a flat verdant valley, in the centre of which, some
time before noon, we reached the pretty " villa"
or town of Santa Barbara. In this vale I saw for

* The Barometer at this place stood at 29·652. Thermometer 86°.

the first time, in Mexico, bright green fresh-looking herbage, as verdant as that of our English fields.
Nothing could be more striking than the change
perceptible in one morning's ride over the mountains,—on the other side of which, the whole way
from the sea coast, the grasses were of the colour of
blighted corn.

We stopped here for a time, at two small huts
within an inclosure, and I stretched myself under
the shade of the two first orange-trees I had met
with, from which in a few minutes a kind woman
brewed me a glass of cool orangeade. This she accompanied with such good stewed meat and white
thin Tortillas, that I began to imagine myself once
more in a civilized country. I went out afterwards
to procure shoes for my mule, and passed through
the town, in which there are few buildings with any
pretensions to be called houses; but the huts are
many of them situated within slight inclosures,
where small fresh-looking trees were growing. In
the Plaza are two of a magnificent size, with immense heads and thick bright leaves, and beneath

their shade a number of country-people were holding a kind of Sunday's market.

The church, which externally was but half completed and looked like a ruin, had three bells on
its tower: and I may here remark, that subsequently
I observed many other village churches of Spanish
architecture, of which the exterior had never been
completed, although within was to be seen all the
glare of Catholic finery.

At five I started again to join the cargo mules,
and rode four or five miles over a grassy plain,
which near the mountains bore marks of the
plough, and the recent gathering in of the maize
harvest. In order to cut off three leagues of the
Camino Real, by which the loaded mules were to
travel, we crossed a small mountain named "Sier-
rita de los Muertos;" and how we reached the other
side without having killed ourselves and animals,
or broken any limbs, is to me quite marvellous: I
never saw so bad a path,—amongst immense fragments of rocks, and deep holes worn by the rains
between them. I would strongly advise all tra-

vellers who have any regard for their own necks, or compassion for their poor mules, to take the longer route, rather than attempt this place by the light only of a rising moon, as we foolishly did. This was once, if report may be credited, the regular Indian road of communication; and might have done very well for such active people as the ancient inhabitants, or the goats, but is not fit to be traversed by horsemen or laden mules. Stumbling our way over a stony descent, we crossed a cultivated plain for about three or four miles, and reached a few huts named " La Laja," inhabited by the labradores of the maize-grounds.

We here slept in a little shed, without knowing where the cargoes had stopped in the mountains.

I heard of them however, on the following morning (the 22nd), as I sat on a stone eating my breakfast of hot Tortillas,—a general article of sustenance in this country*, which I may endeavour to describe before the mules come up.

---

* The maize was prepared in precisely the same manner before the Conquest. See Bernal Diaz, and others.

The maize of which the Tortillas are composed is first parboiled, to cleanse and soften the grain, and then, in a quantity sufficient for the day's consumption, is left to cool.

For the purpose of crushing or mashing the maize, the women have a large square block of black lava or basalt, about two feet in length and sixteen inches broad, which stands on two, three, or four legs, so arranged as to give it a gentle slope. There is a very slightly elevated rim on either side, and the great solidity and weight keep the stone steady, while the operator bruises the maize with a long stone, not unlike a rolling-pin, which is held at each end, and so moved that it crushes the grain to paste, and at the same time pushes it down to a bowl placed ready to receive it.

This process is gone through once, twice, or more, according to the fineness required; and where great care is taken, it is passed through a fine sieve. A lump of this paste is then taken and patted skilfully between the hands until it becomes as thin as a light pancake; and the great art consists in thus flattening it out without breaking the edges. The cake is then laid on a smooth plate of iron or flat earthenware, which is placed over some charcoal or wood embers, and kept at a certain heat: here, first one and then the other side of the Tortilla receives a toasting, and great care is taken that it should not be at all browned. The grand object in the latter part of the process is to serve up the Tortillas hot and hot, as fast as possible, in a clean napkin; and a slow eater who begins his first Tortilla, will find twenty or thirty piled in a smoking heap at his elbow long before he has made any progress with his dinner.

The making of Tortillas is so important an art, that in the houses of respectable people a woman, called from her office " Tortillera," is kept for this express purpose; and it sounds very oddly to the ear of a stranger during meal times, to hear the rapid patting and slapping which goes forward in the cooking-place until all demands are satisfied.

From the Rancho of La Laja we began the ascent of the Cuesta del Contactero, an immensely

precipitous mountain thickly covered with wood;
and having ridden two hours over a terrible road,
we reached the summit, on which is a cave containing a most miraculous picture of the Blessed Virgin, in the side of a steep rock. The cave is about
the size of a sentry-box, although somewhat higher,
and is shaded by magnificent oaks: it is composed
of a light-coloured stalactite, with which all the
surrounding limestone rocks are thickly coated;
and the projections or irregularities have formed
small niches and crevices, which a lively faith has
metamorphosed into a representation of Holy subjects. A large shapeless hollow in the top of the
cave is considered as the Glory over a figure of
Maria Santissima, which good Catholics see most
clearly delineated on the smooth back of the cave.
I confess that my heretical eyes saw nothing but
several of those irregularly waving lines which are
perceptible in most rocks of this kind; but being
directed by a border of tawdry paper and tinsel
roses, I was enabled to discover something like
the outline of a large sack.   To this my Arrieros

climbed up and paid most profound adoration; pointing out to me afterwards, the face, eyes, mouth, and beauties of the Blessed figure, in whose arms they also discerned the Holy Child as clearly as possible!—The cave stands at an elevation above the road of about twenty-five feet, and is ascended by climbing up the rocks, now worn in many places as smooth as glass by the devout persons who have ascended to pay their adoration.

" The Holy picture of Maria Santissima de Santa Barbara was discovered many hundred years ago by a priest, who when travelling found a number of Indians performing their idolatrous ceremonies on the summit of this mountain ;—a circumstance which so fired his holy zeal, that he determined on the reformation of these Pagans by a miracle. Seeing a quantity of bushes growing out of this cave, (in which it is certain no bushes could ever have grown,) he ordered his Christian attendants to fire them; and when their flame was extinguished, the figure of the Mother of God was seen stamped, as it is to this day, on the solid rock."

—A general baptism was the natural result of such a manifestation; and the good Padres of Santa Barbara have ever since derived a comfortable revenue from the miracle.

From this sanctified spot, in about three miles we gradually descended to the delightful valley of Los Gallitos, embosomed in the Cordilleras, and clothed with a bright short sward, on which a quantity of fine cattle were feeding. It is a little space, three or four miles in length and about half as broad; and through its centre, amongst pic-turesque irregularities of ground, flow two small clear streams as they descend from the surrounding mountains which are thickly clothed to their sum-mits with magnificent oaks. Several spaces in the valley are fenced off and ploughed; and the snug cottages, with domestic animals wandering near them, reminded me forcibly of the scenery in some of the small Swiss vales.

Pitching the tent, we lay down in a temperature of 82°, to sleep on the fresh soft grass. A little rain fell during the night, and the Rancheros in-

formed me that showers towards midnight occurred almost constantly; so that a perpetual verdure reigns in this charming place. This valley,
with another equal to it in richness and beauty
and a short space to the left, belongs to the Carmelite Friars of San Luis, who are also the possessors of a variety of others of the most fruitful
spots in this part of the country.

May 23.—We began the ascent of the Cerro de
los Gallos very early in the morning, that our
animals in this laborious journey might not be
exposed to the full heat of the day. In an hour
and a half the saddle-mules reached the summit,
and in another hour the cargoes joined us*.

On the mountain top we found blackberries in
ripe fruit and flower, and the wild geranium of our
English hedges abounded; oaks, which were the
only trees, formed a noble forest. Hence we descended two miles to a little green plain, and breakfasted at a solitary hut, after which the road again

* The Barometer here stood at 25·750. Temperature 65°;
while in the valley it was 27·156. Temperature 79°.

became exceedingly bad and stony, amongst rugged hills for about nine miles, when by a long
rocky descent we at last reached the lower country, arid and dusty, cut up by the deep irregular
channels of mountain torrents, and nearly bare of
trees*. In many places it had been extensively
ploughed, and was lying fallow. On our right, at
a mile distance from the road, was a lake of two
miles in circumference, which is said to be quite
salt, and no living creatures are found in its waters.
At about half a mile to the left is the ruin of an
Aztec building, which I sketched, and then ascended. The base, stretching in diameter about thirty
paces, rises in the form of a portion of a cone; it is
very neatly built, and has been entirely faced with

* On the burning gravelly soil I caught an animal of the
Lizard tribe, which is called "Cameleon" by the natives. It
was about the size of the palm of the hand, and nearly of the
form of a turtle's shell. Its skin was coarse, of a dingy yellow barred with dusky brown, and a number of rows of
fleshy prickles or points ran longitudinally over its back and
sides. The tail was very short; the whole creature flat, and
so lethargic as not to attempt its escape. This animal is figured by Clavigero under the name of "Tepajaxin."

flat unhewn stones, and has others projecting at intervals, as if for steps. This part has been much defaced by the Indians, who live in a hut near it, having carried away the stones to build a wall for a cattle yard; enough however remain to show what it has been. The base rises about twenty feet, and on its summit is a kind of walk or parapet round a small tower, which is now only about ten feet higher; and the whole edifice is filled with hard earth or clay, in which the Indians have found several stone and earthenware figures: some of the latter of these, although much mutilated, I purchased of their children. I should conceive, from the shape of this building, that it must either have been a Teocalle (or temple) or some very large altar of sacrifice, since several smaller ones, which resemble it in form but are only three or four feet in height, are still said to exist in the neighbourhood.

Hence we rode over the dull, dusty, and most uninteresting country, bearing here and there a stunted bush or a Yucca-tree, and in eight miles reached

Tula, (I believe a city,) which with the exception of the Plaza, where are a few ruinous houses, is composed of huts built of large bricks of unburnt clay, called Adobes. This, with the dust, the palm-trees and the pulque* (which in its sweet and fermented state also has precisely the same smell and flavour as the Lackbi† of Africa), reminded me most forcibly of Morzouk, the capital of Fezzan, as I rode up the broad dusty street of huts, under the glare of a scorching sun, to a Meson in the Plaza, where we procured a bare room. Our lodging was opposite the church, at which, hearing music in the evening, I found a crowd of people with a young woman who was bearing on her head a little dead child, dressed in coloured papers so arranged as to represent a robe, and tied to a board by a white handkerchief. Round the body were stuck a profusion of artificial flowers; the face was un-

* This is the first place at which I found the *Agave Americana* or Maguey cultivated for the purpose of extracting Pulque; of which as there is a great consumption, I soon found there was also a great deal of drunkenness.

† Inspissated juice of the Date-palm.

covered, and the little hands tied together as if in prayer. A fiddler and a man playing on a guitar accompanied the crowd to the church-door; and the mother having entered for a few minutes again appeared with her child, and walked off, accompanied by her friends, to the burying-place.

The father followed with another man, who assisted him with a lighted piece of wood in throwing up hand-rockets, of which he bore a large bundle under his arm. The whole ceremony was one of cheerfulness and gaiety, since all children who die young are supposed to escape purgatory and to become " Angelitos * " at once. I was informed that the burial would be followed by a fandango, in token of rejoicing that the babe had been taken from this world.—It is doubtless the duty of Christians to be resigned to their afflictions; but I am sure that few English women could carry their first and only infant to its grave, with smiling countenances; and I equally can

* Little angels.

answer for the inability of the men to throw up rejoicing rockets when their first-born is taken from them. I entered the church, which was neat, and, according to custom, crowded with images: before one of these a sallow wretched man was kneeling, with his arms extended for so long a time that it became painful to look upon him, and I left him to perform his agonizing penance ungazed upon by the eye of curiosity; since whatever may be the errors of the creed which imposes bodily suffering as an atonement for sin, his was an act of fervent piety, and, as such, was sacred. Hence I visited a school, attracted by the noise of seventeen little boys repeating at the full stretch of their variously toned throats the " Ordinanzas" of the Church, each one bawling with all his zeal and with all his strength. Their master, a fat, lazy, good-tempered-looking man, fairly lost his patience in endeavouring to make me hear, through the din, his questions as to " whether the Spaniards would come again and hang all the revolutionists;" and soon gave them their dismissal, after they had knelt down and rapidly screamed out two prayers which he named to them: each child then came and inclined his shoulder to receive the blessings of the master and the stranger, and after this very pretty little ceremony they all ran off whooping and hallooing down the street. I saw but one book in this extraordinary seat of learning; but the master very seriously assured me, that several of the boys could nevertheless read.

On my return I found a white well-dressed lady of Tula, with three very pretty children, in a sad passion with my servant, who had turned a deaf ear to her offers of purchasing one of my three English fowls, of which I wished to carry the breed to the interior. No excuses would satisfy her; and on my appearance to confirm the refusal of two Tula chickens for one English, she indignantly brushed out of the room, declaring that she had been rightly informed that " the English were ill-natured animals." This well-bred woman was the most considerable person in the place, and my hostess was all astonishment at the man who could refuse her any thing. I slept in a wretched room, for which I paid an exorbitant price, and had, besides, the misery of being obliged to attend to the complaints of a long list of ladies, all either pregnant or hoping to be so. My landlady had spread a report of my being a "cunning man," from my having told her some extraordinary truths (of which her husband had previously informed me), as to her name, her being thirty-three years of age, &c.; and my having satisfied her anxious inquiries as to the sex of her next child, which I informed her "might possibly be a boy." I have already remarked how much Tula resembled Morzouk; but these circumstances confirmed the comparison, since I here had to answer precisely the same questions as I had been asked by the African women, to females not at all wiser than those poor despised Negroes. My own country-women would scarcely credit some of the anecdotes I could tell of those neglected creatures in the northern states of this Republic; and would not believe it possible that the females of any country called civilized, could be so barbarously kept in a state of the most perfect ignorance.

May 24.—Early on the 24th we set out over a parched plain, studded with dwarf mimosas, palms, and tunas. At about twelve miles we pass-ed a few huts, the Rancho de la Boréga, on our right, and on the left of the road saw a venerable-looking old man sitting beneath the shade of an acacia, in the branches of which, immediately over his silvery head, a crucifix was placed. Imagining that this picturesque personage was a hermit, I rode up to salute him; when he lifted a cloth which hung over a bough, and displayed to me some bowls of " Atole de leche," a composition of the finer and more glutinous parts of maize, boiled milk, sugar and cinnamon, equal to any of our custards. We each took a bowl, and discussed its contents with spoons made of palm leaves, completing the meal by the purchase of a quantity of the fruit of the tuna, or prickly pear, with which several laden asses were proceeding to

Tula. An Indian from the interior passed us here, carrying his bow and a handful of light arrows headed with small narrow pieces of copper.

Riding forward about ten miles, we passed over a rising ground entirely covered with a stream of lava *, which in large irregular masses appeared to have issued from two slight elevations on the left, bearing all the aspect of having once formed part of a volcanic crater. The hills immediately on the right were of limestone.

Not far off Flores showed me a rock, of which there runs a tradition that " an old very strong necromantic Indian, flying with his treasure at the time of the Conquest, brought it to this place, and by his strength and art overturning the stone, buried his riches beneath it. They still lie there, and can only be obtained by the stone being again overset by any number of men under eight, which number destroys the charm and the gold is lost: seven Arrieros had re-

* The name of "Mal Pays" is applied throughout the country to places of this nature.

peatedly tried their powers, but without effect."
All the space round the stone had actually been
scraped up by the credulous; but I observed that
these efforts to overturn it had been made in one
direction, at which a projecting point was the impediment.　Digging beneath this, we three Englishmen assisted by Flores, tumbled the magic rock
over with such force that it split in two, but found
no treasure to repay us for our trouble.

Flores told the story triumphantly to all we met,
and who seemed acquainted with the rock, which I
lamented having broken, since a favourite object of
Arriero tradition had thereby lost its charm.

During the greater part of our ride we had
been accompanied by a patient of mine,—a mule-
teer, whom I had cured of fever and ague; with
his master, a very drunken old Ranchero, who
talked politics and drank Vino Mescal (a liquor extracted from a species of maguey,—and re-extract-
ed by him from a bottle which hung at his saddle-
bow,) until he became very troublesome and tormenting.　This man, the owner of a large estate,

was one of many I had recently met who were in
sad alarm that Ferdinand the VIIth was coming
over to cut the throat (and here he drew his hand
across his neck) of every Mexican. He also with
the multitude could not be persuaded that Spain
and Europe were not two names for the same
country; London he believed to be a nation of
itself, England an inferior province, but both occupying remote corners of the king of Spain's do-
minions. As for France, he affirmed that it was
but another name for Panama.

At six miles from Mal Pays we arrived at the
Rancho del Coronel, having ridden the whole way
in one unceasing cloud of dust. In a hut here I
found two merry Silenus-looking old fellows on a
large bull's hide, sitting in high glee over an immense jar of pulque, very drunk and tormentingly
polite to myself and each other; lavishing all the
compliments of the country on their own acknowledged, and my supposed merits. Happily for me,
the mules were able to pass on to La Viga, two
leagues from " el Coronel."

From hence I sent forward one of my people to Mr. Price, my medical friend, who was two days in advance. The man had long fancied that each dusty respiration was to be his last; and without any serious illness, was determined on killing himself by the mere force of imagination. We pitched our tent on a barren plain; and when the sun went down found the night so cold, that we were glad to warm ourselves by drawing water for our cattle by the rude machinery placed over a well of bad water ninety feet in depth.

May 25.—At daylight on the 25th the thermometer stood at 65°, which after the heat of the low country appeared almost a freezing temperature. Our day's ride of thirty miles was over dull uninteresting ground, covered at intervals with mimosas, which by natural outlets yield a great quantity of fine transparent gum. The maguey was very abundant as far as the eye could reach, and dust and whirlwinds half-blinded us. At about twenty-five miles we came to a very stony hill, on which were standing about three dozen huts named " Buena Vista;" and in five miles more we reached the Hacienda del Quelintal, at the foot of a similar hill, where I joined Mr. Price. Here the water, which can be procured only from wells thirty-five feet in depth, is harsh and unpleasant. We took up our abode in the hut of a very poor man, and at night lay outside beneath the eaves, while near us reposed our host, his wife, and eleven children, forming a line of no small extent; each little child was rolled in the ragged blanket which constituted its clothing by day, and the air of contentment which reigned throughout this miserably poor but uncomplaining family, would have taught a wholesome lesson to a dissatisfied man. This was not the only time I had occasion to admire the mild unrepining manners of a class of the Mexican population, of whose poverty no idea can be formed but by eye-witnesses.

Our host paid two dollars per annum as ground-rent, and three more for a sufficient space in which to sow half a fanega of maize, on the produce of which the family in great measure depended. I wished that the poor fellow might have some pulque to cheer his heart; but his landlord required five dollars a hundred for the magueys, although thousands and tens of thousands were growing wild and decaying all over the plain *.

May 26.—We left Quelintal at seven, and I had now the satisfaction of having Mr. Price for my fellow-traveller during the remainder of the journey. The road lay over several low sierras and rugged paths for four leagues, until we reached a snug cluster of huts called Rincón. Here we found a sick man in a raging fever, to whom his friends were administering a rich stew of blood and sweet herbs. He lay moaning in a corner of our room, across which a rope was stretched, supporting four quarters of recently killed beef, with other morsels hanging dangling and bloody. A fat woman very drunk sat on the threshold of the door; while another still fatter, and but little more sober, took her station on a bed close to our table, from whence she helped herself to our dinner and

* The Barometer here stood at 26·300. Thermometer 79°.

whatever else she fancied; taking possession at length of our jar of pulque, which she shared liberally with her friend at the door. We found it impossible to escape from these two Bacchantes, our tent being like an oven *, and were therefore obliged to exercise our patience until the sun went down.

In consequence of Mr. Price having bled the sick man, we were sadly tormented by patients, real and imaginary, under the immediate chaperonage of the fattest old woman, who finished the day by coming with some more invalids into our tent after dark, squatting her spacious person close to my face as I lay in bed, thereby breaking the sacking, and in consequence receiving from my naked foot an impetus which sent her in an instant through the tent door.

May 27.—The thermometer at dawn was 68°. All our patients were better, although some little discontent was shown at our not having instant

* At two P. M. Thermometer 88° in the shade; in the sun 118°, and in the tent 104°. The heat of the ground was 137°.

remedies for the blind, the fat, and the lean; palsy, pregnancy, and its reverse. At seven we travelled over a series of bad roads amongst the mountains; and at ten miles passed the Rancho of San Isidro, in which was the only cottage I had ever met with where any attention had been paid to exterior appearance; some climbing plants were clustering over its front, and from amongst them peeped the two first roses I had seen in blossom. In twelve miles more we reached the huts of Puerto San José *, in a dell amidst the mountains, where, under a fine shady tree, a gay party were assembling to set out on horseback to a wedding at Peotillas. My servant Marriot was here attacked with strong fever, but was much relieved by copious bleeding.

In the evening I bathed in one of the troughs at which the cattle are watered, and found the temperature of the well (67°) so cold in comparison with the air (84°), that on plunging into the water the circulation left my fingers. While dress-

* Barometer 25·330.   Thermometer 83°.

ing I suddenly heard a loud bleating, and in an instant above four hundred milk-white goats sprang down the side of the steep little ravine, and rushed to a well at which young shepherds and girls were drawing water for their evening's drinking. The stillness and clearness of the star-light evening, the picturesque dresses of the shepherds and young women, each with their favourite kid, re-called to the mind the description of the Patriar-chal Ages, when the maidens attended their fathers' flocks, and assembled round the wells with the shepherds as the sun declined. We slept under a mimosa near the fires of our muleteers, with whom on the morning of the 28th we had much trouble, as their mules were tired and exhausted, and none others could be hired to supply their places.

In three leagues over a stony road we reached the Rancho del Tejou, and passed on to a plain on which the number of whirlwinds was quite ex-traordinary. We had repeatedly seen a few of them, but on this day they appeared to have as-sumed a new form, raising the dust to a height of two or three hundred feet in straight columns which preserved their perpendicularity and moved but slowly over the plain, while many continued to turn rapidly on their axes without any perceptible progressive motion.

After riding three leagues further we reached the plain of Peotillas, where the younger Mina, having but 172 men, including himself and staff, so gallantly defeated Armiñan with a force of 1700 disciplined troops *. The Spaniards were nearly all destroyed, and we visited the long grave in which many of them were buried, beneath a rude cross of wood, whereon was engraved

> *Vn Padre N^{ro} y un Abe Maria Gloriado*
> *—y un Sudario Por Yntinsion de las Ani-*
> *mas q^e estan Sepultadas en este Campo.*

The field of battle was covered with low bushes,

* " On the body of a lieutenant-colonel was found the order of the day, which showed that the force actually engaged, was six hundred and eighty infantry of the European regiments of Estremadura and America, and eleven hundred of the Rio Verde and Sierra Gorda cavalry, and that the rear guard consisted of three hundred men."

and of course well calculated to favour a force such as Mina's, which consisted for the greater part of raw undisciplined Indians and country-people, who in this covered ground might be disposed to better advantage than a more regular body of men.

I was extremely interested by the narrative of Mina's campaign, written with great feeling by Mr. Robinson, and full of the interesting incidents which occurred to that brave but ill-fated man, during his short but astonishing career in Mexico. One league from the field is the Hacienda of Peotillas, situated at the foot of a low rugged chain of hills. There are a few good dwellings here, with a church and store, of stone. In the portico of the Administrador were hanging two stuffed wolves, and a neighbouring tree was gar-nished with the bodies of numerous coyotes, the jackals of the country. I was fortunate in visiting the Hacienda at the time of watering the horses and selecting some for sale. Above three hundred were careering about in a long inclosed space, with several Vaqueros on horseback and on foot attending them with their lasos.

Wishing to purchase a horse, I was enabled to see the laso exercised in its utmost perfection; and the dexterity with which particular animals were selected and arrested for my inspection from the herd at full speed, was far beyond what I had expected. The whole scene was of the most animating kind; the wild galloping horses, the mirth and activity of the men on foot, who seemed delighted in showing their skill, was above all things pleasing. I purchased here a very good pacing horse for twelve dollars, equal to 2*l.* 8*s.*, and then rode on to a small tank at about a mile distance, where for the first time in my journey I was refused admittance into a hut until the tent arrived, although a gale was blowing and we were half smothered with dust and gravel. Our road this day had been so uninteresting, that I now gave up in despair all hope of agreeable travelling on this part of the so much vaunted " Table Land of Anahuac."

What with the excessive heat and monotonous
surface of the Tierra Caliente, the difficult and
fatiguing ascents of the mountains, and the clouds
of dust of the " Temperate Regions;" I began to
be rather tired of my journeying on a road so
totally destitute of interest or incident. My chests
and furniture were split by the sun, or by the
laden mules knocking them against the trees; and
instead of being twelve days, as was expected, we
had now been travelling fifteen. Fifty miles of
our journey yet remained to be performed, with
jaded mules, and Arrieros sick with ague and fever.
I do not complain either of my food or lodging,
being always grateful for them whether good or
bad; but fastidious persons would do well never
to enter the Mexican territories *viâ* Tampico and
San Luis. It should be the constant axiom with
the stranger, that whatever feeds or covers the
people amongst whom he travels, will unquestionably nourish and shelter himself; and on this
principle, he will find no difficulties in earthen
floors, in mud huts, tortillas, or ropes of beef. In

VOL. I.

the afternoon my sick servant was again bled, and I decided on leaving him to the kind care of Mr. Price and riding into San Luis to prepare lodgings for their reception, leaving them to divide the journey into two days.

May 29.—At daylight therefore on the 29th I set forward with Flores, in a temperature of 50°, which by noon rose to 84°. In seven miles we passed a few huts called La Colorada, and in ten more reached a solitary hovel near the road, where Flores was taken very ill, and I bled him copiously, after which and a few hours sleep he was sufficiently relieved to go on again. A day had scarcely passed since leaving the coast without my having met with some persons labouring under sharp fevers and agues, which it appears are not confined to the low country, but, on the contrary, become rather more frequent on the Table Land, notwithstanding its reputed salubrity. In twenty-six miles more we entered the town of San Luis at sunset, soon finding our kind agent Mr. Dall, who obligingly invited me to his house. Our road, as usual, had been desolate and uninteresting, with few trees, but great abundance of gigantic nopals, the trunks of which were many feet in diameter. Hares and rabbits were very numerous, and so tame as scarcely to leave the road when we passed them.

At about seven or eight miles from the town the irrigated gardens and fields commenced, and broad straight roads, covered with dust to above a horse's fetlocks, crossed them in all directions. In the outskirts, Flores pointed out to me two graves containing a number of Spaniards who had been killed in the war, and on which a singular figure was scratched or scraped every night by the unearthly hand (as was generally believed) of a soul in purgatory. The priests, with their usual anxiety to prevent imposition, are said to have set a watch to detect the nocturnal limner, but without effect; for as we passed, the mysterious mark was still visible. Not far from the place is a small building with holes like those in a post-office window, into which a number of inscriptions invite all good

Christians to put their money to purchase masses
for the Almas de Purgatorio (souls in purgatory), .
which to heretical eyes soon discovers the mysterious nocturnal artist.

May 30th.—I rambled out early in the morning to see the market of which I had heard so
much; and was not disappointed as to its neatness and novelty, although I did not find such
abundance or variety of fruits and vegetables as
I had been led to expect. It is held in a large
square, and presents to a European an appearance so unlike anything to be met with in the Old
World, as to be very interesting. On entering,
I passed a long line of Indian and other women
squatted on the ground and selling roses and
nosegays. Some dealt in fruits, others had coloured pulques in glasses; and there were not a few
occupying themselves over little charcoal fires,
preparing cakes or frying meat for the consumption of the country-people, Arrieros and Leperos,
who swarm in the street and market. Further on,
the square was filled by little spaces ranged in

lines and shaded by a mat or two, erected on a
pole in the form of an umbrella, or by some rude
kind of awning on four sticks. Here peppers,
legumes and fruits, are spread upon mats on the
ground for sale, while the venders sit cross-legged
or squatted by their goods,—some indolently lounging and smoking their cigars; others dozing half
asleep, and wrapped in their variously coloured
Serapes. In most instances each stall merely appeared to contain the produce of the gardens of
those who attended it, for many did not at first
contain a dollar's worth of goods or fruits; and
when these were sold, a new comer was at liberty
to occupy the stall. On the opposite side however
to that for the fruits and vegetables are many
larger standings, in which glass, crockery, cutlery,
articles of leather, bitts, &c. are exposed for sale,
always at double, sometimes at treble the prices
for which they are ultimately sold. On one side
of the square is a long colonnade or piazza in
front of a large public building called " Alhondi-
ga," which consists of a quantity of spacious stores

for corn and merchandize, and is of great extent. In the colonnade are a number of little temporary stalls, at which very inferior European hardware, trinkets, looking-glasses, and other articles, are sold at an exorbitant price. The meat market is not in the same square as that for other goods, but in a short broad street running out from one corner of it. The flesh is in large coarse dirty-looking lumps, either baking in the sun or co-vered with flies; and the shambles suffer much by comparison with the other market. In addition to the fixed traders in these places, a number of hawkers are constantly to be seen mingled with the crowd, and offering to the by-standers boots, blan-kets, skins, whips; in short, every variety of pro-duce, and all selling on commission for the ma-nufacturers, to whom a certain sum is returned,— the broker appropriating as much as he can, to cover his own expenses of time and lungs. In passing through the streets many cages of birds are seen suspended within the wooden barred windows, of which the sinsontli, or mocking-bird, seems to be the favourite. These are very abundant in the surrounding country, yet fifty dollars are considered as by no means a high price for a good singer. In the afternoon I waited on the governor with my passport, and found him to be a plain well-bred old man, apparently much respected *. Notwithstanding his situation, he is a shop-keeper, and sells, or causes to be sold for him, all that variety of goods which may be found jumbled together in a Mexican store. This occupation is by no means considered a degradation, since there are few persons of consequence in the place who are not in the same circumstances; and many of the great men of the land actually keep drinking-houses. In returning home, the tinkling of a bell attracted my attention, and every one around me instantly fell on their knees: a carriage with two piebald mules had just drawn up before the cathedral door, and a priest in his robes enter-

* On the top of the Government-house, I saw for the first time the cap of liberty carved in wood and painted red, surmounting a large flag-staff.

ed it, followed by two choristers bearing censers after the Host, which he was conveying to some dying person. The guard turned out under arms, knelt and grounded their pieces and pulled off their caps, while the priest who carried the Host leant back comfortably in the coach and treated himself to a pinch of snuff. There are more kneelings and pulling off of hats at San Luis than at other places. No one passes the door of the Parroquia, or cathedral, without making a reverence bare-headed; and many good Christians perform the same marks of respect to the other churches. This obeisance is exacted from no one; but were a stranger to omit it, he would stand a very good chance of being insulted by the very bigoted populace, who still view with a jealous eye the heretical foreigners who have lately appeared amongst them.

At certain hours the bells of the Parroquia announce that all should be uncovered, and that a prayer should be recited, a rapid tinkling giving notice when the hats may be put on again. I la-ment much that I was not in San Luis at the time of the ceremonies on the day of Corpus Christi, having heard that they were the most splendid exhibitions during the year. One of the images in this procession represented the Supreme Being bearing the Saviour in his lap.

In the evening I went to Flores' house to give him some medicines, as he was still very ill; he lived immediately opposite the Cock-pit, which I visited and found crowded to excess by all conditions of men, from the best-dressed persons to those who had but a portion of ragged blanket to cover them, but each carrying his own bird. The pit is a circular space, surrounded by seats of masonry and shaded by canvass; no money was paid for admittance. I saw much betting, and several stakes delivered into the hands of a kind of master of the ceremonies; but I did not remain to witness a fight which presently took place, and from which in less than five minutes three dead cocks were brought out. Bull-fights had recently been prohibited, but I believe much to the regret of the people. I visited the

Plaza de Toros, in which it has been the custom to hold a kind of wandering evening market, where all sorts of wearing apparel and other articles are hawked about.   There are but few shops in San Luis at which any of these conveniences can be purchased, and it is therefore the usual custom to send them out to look for customers. Here also stolen goods are not unfrequently sold; but from the general use of the knife amongst the " valiant sons of Anahuac," I should be very unwilling to arrest a thief in the crowd of wild ill-looking fellows who frequent the evening mart. While each vender was auctioneering for himself, and a second Babel seemed to reign, the vesper bell tolled.   A universal silence prevailed,—every head was bared, and all business was at a stand. At the next tinkle, bidding, cheating and quarrelling, were as instantly resumed.   At dusk I again met the Host, preceded by men bearing lanthorns, on its return from a dying person.

The following story, which was related to me by several gentlemen of strict probity, serves to show the present state of ignorance in medical science in San Luis Potosi. A physician, who was pointed out to me, was called to the assistance of a poor labourer with a ruptured blood-vessel. Ice was the only remedy known to stop the flow of blood, and none could be obtained until a priest should be sent for to confess the sufferer before he died. What then was to be done?—it was but too certain that the man would soon expire, unless means were found to arrest the effusion. The physician therefore had no resource but to sew up the poor wretch's mouth and nostrils; but even before the Host could be sent for, the miserable creature was, very naturally, suffocated.

May 31.—I visited the church of San Francisco, with the monastery attached to it, both handsome buildings, and according to report richly endowed. We were first ushered into a large court having a grass-plat in the centre, round which ran a corridor; and on the walls were hung a series of pictures descriptive of the life and actions of San Francisco. The saint is hungry, and our Saviour is seen sitting at table with and helping him to the viands, while an angel brings him water to quench his thirst. He is represented, after his death, as sitting on the left hand of the Almighty, who is pictured as an aged man having on his right the Redeemer and the Virgin Mary. Some of the supposed miracles and conferences with the Divinity are beyond all description blasphemous; and the Friars informed me,—at the same time triumphantly pointing to a painting of the event,—that the saint had procured from God's own mouth the entire pardon of the sins of the world, as fully as was granted to Christ himself; but that the *Pope* of the age in which San Francisco lived, would not confirm the grant, and that mankind in consequence had been left in their blindness and sin!

I saw also a gorgeous painting of Christ transferring the marks of the cross from his own person to that of the saint, before the holy man was carried by angels to heaven. In an upper gallery the life of San Antonio is portrayed in a series of miracles as profane as those of his brother saint below.

I should tire and shock my readers were I to tell them all; but one subject is so novel that I may be forgiven for describing it. This is the figure of an ass kneeling, with its mouth open, before the saint who holds the sacred wafer, and who has just had recourse to this miracle to convince a number of heretics who had doubted the actual presence. The interior of the church is handsome and airy, and in some places tastefully ornamented by good carved fretwork in stone. The greater part of the statues were of monkish saints in the cloth robes of their various orders, which had a far better appearance than the gaudy dresses and tinsel of the usual figures. On the whole, there was good taste in the decorations of the church, and in one aisle is a chapel "most particularly blest (de un modo muy singular) by the Divine Presence," in which is a temple-shaped altar, very prettily executed in white and gold, and containing the only rational-looking statue of the Virgin I ever saw.

There were no spangled hoop petticoats, crowns, wigs, or artificial jewels; but a handsome woman of the natural size, dressed in a robe of unglazed black silk, the folds of which hang gracefully over one arm. The bust is well proportioned, and covered to the throat, and the waist is encircled by a plain gold band. The contrast between this figure and the simple white temple is very striking, on account of the extreme plainness of the whole. There is but one good painting in the church, and this is a copy of the Divino Rostro (Divine Countenance) by a self-taught artist and architect, Señor Tresguerra. It is merely a representation of the head of our Saviour on a sheet, but beautifully executed; and presented by the artist to the church, on condition of its supplying the place of a very ugly statue.

I had waited with anxiety Mr. Price's arrival with my sick servant, but received a letter from a place called " the Adobes," seven leagues distant, saying that he was unable to proceed further on that day; which was greatly to be lamented, as not

a single comfort could be procured for the invalid:
even the water, which was obtained from a tank,
being thick and undrinkable.

June 1.—With much difficulty I hired a carriage to proceed on the following morning for my
poor man. In the forenoon, being the eighth from
the celebration of Corpus Christi, a temporary
altar was erected in a small lower room at the
corner of the House of Congress, or Palaçio, which
faces the Parroquia. From thence at ten o'clock
a procession sallied forth and performed the cir-
cuit of the Square, which was crowded with people. Mass was read to the kneeling multitude; and
the procession then returned in due form to the
church, preceded by a number of ragged Indians
and half-casts, playing on rude fiddles, guitars, and
flutes, each according to his own particular fancy
or ability. Then followed a confused mass of women, some carrying candles, others flowers, and
four of them bearing a statue of San Cristoforo,
who appeared to be some Catholic Hercules, he
being dressed very like the ogre in Tom Thumb,

and wearing an enormous sword buckled round him with a broad buff belt. The platform on which he stood was ornamented with ears of ripe Indian corn, apples, flowers, and tinsel; and I observed that there was a great competition among the women for the honour of bearing this doughty saint. Next followed the Host, borne by a priest under a gaudy silk canopy, and attended by others in their robes of ceremony, each with his hands uplifted and devoutly pressed together, while their eyes wandered about the crowd in search of acquaintance, whom they acknowledged by a nod. The members of the Provincial Senate in full suits of black, and bare-headed, closed the procession. Don Francisco Tresguerra,—the artist whose "Divino Rostro" I had so much admired,—was so obliging as to visit me. I was quite charmed with the good sense and taste of this gentleman, who has devoted his talents, I believe quite without benefit to himself, to the embellishment of San Luis, where some of his works, both in painting and architecture, reflect the highest credit on his abilities. He was at this time building a theatre at one extreme end of the town, an altar-piece at the other; and painting besides, on a very large scale, for three or four different churches: he has never left the Republic, and from his age, now about fifty-five, it is not probable that he will ever gratify his prevailing desire to visit Europe. The exterior architecture of the churches of San Luis is generally very heavy, overloaded with carved ornaments and ill-executed statues of saints; yet at a short distance they give a magnificent appearance to the town, which contains many very large buildings. The Palaçio, now the Provincial House of Congress, is of this number, and forms one side of the Plaza de las Armas, which has an excellent fountain of water in the centre. The Parroquia occupies a portion of the opposite side, and on its right are the soldiers' quarters. The two other sides are filled with shops and dwelling-houses, it being the custom for the owners of good mansions to let the lower story as a shop. Mr. Dall's store was in the lower part of what was once the Custom-house, but now is the dwelling of a rich widow, one of the most important persons in the city. She kept her carriage and pair of mules, and was a great lover of the fine arts, as might be seen by the very extraordinary collection of little carved figures and toys nailed round the door of her Sala. Amongst the large public buildings is a very salutary one for the confinement of refractory women,—jealous fathers or husbands enjoying the privilege of shutting up their daughters or wives! The church attached to this virtue-guarding edifice is very dark and gloomy, and grated off at one end like a nunnery, so that the poor women cannot be seen, although they themselves are able to peep as much as they please. The large house in which the Inquisition was formerly held is now a private dwelling, and does not possess that gloomy aspect which one expects to find in such a place. The Meson Nuevo (New Inn) is a neat clean house, the cooking good, and the people civil and obliging: but in the rooms there is no furniture of any kind; and the rent, stabling, and cooking expenses are all on separate accounts. The shops of San Luis are good and well stocked, those for liquors being by far the most numerous. All leather, iron-work, crockery and coarse goods, are sold, as I have already observed, in the Plaza de los Toros. The town is plentifully supplied with water from wells, but there are also regular water-carriers who bear four large earthen jars in a kind of wheelbarrow. Pulque is sold at almost every corner, and its effects on the natives are often very visible. To the abundance of this and other liquors may be attributed the frequent and sanguinary quarrels at this place, and the numerous assassinations committed, chiefly among the lower orders, who all carry knives concealed under their blanket, although the laws formally prohibit weapons. Very slight provocation is sometimes sufficient to make one man stab another; and two murders of this kind took place in open day during my stay at San Luis. The assassin in such cases is merely confined for a few days, and then set at liberty to commit fresh enormities. Sometimes, but rarely, he is sent as a convict for two years to Vera Cruz. One of the murders before alluded to took place in consequence of a dispute between two men of different villages, each of whom claimed for his Pueblo the merit of having sent the largest nosegay as a present to the Virgin at the feast of Corpus Christi;—to end the matter, one very deliberately stabbed the other, wiped his knife, and was taken into custody, well knowing that in a few days he should regain his liberty. To instance still further the state of the laws in the northern parts of the Republic of Mexico at this period, I may relate that a German gentleman some time since was attacked on the road to Durango by a robber, who having fired at and missed him, was instantly shot dead. The gentleman was afterwards fined five hundred dollars for killing instead of taking him prisoner and bringing him to Durango, whence after a few days confinement he would again have been turned out upon the world.—The principal streets are lighted by large lanthorns, yet all who pass take the precaution of going armed, although an express permission must be first obtained to wear a sword. Peaceable men are thus prohibited from carrying weapons of defence, and when killed, their murderers remain unpunished. Three hundred permits to carry arms had been transmitted from Mexico, but they were very insufficient for the principal inhabitants. The military at this time consisted of nearly eight hundred men, of whom about half were infantry, and the others horse artillery, who all appeared in tolerable order, when compared with the wretched rabble I had seen acting as soldiers at Tampico.

June 2.—The coach arrived in the evening with Mr. Price and Marriot, and our invalid was somewhat better, but exceedingly weak. A fresh accession of fever however came on (June 3), and continued unabated until Saturday night, when the poor fellow breathed his last at midnight. In closing the eyes of poor Marriot I lost an invaluable servant and friend, who had followed my good and ill fortunes for six successive years. He had been with me in the brightest as well as the most gloomy days of my existence; and the services on which we had been engaged together, equally attached the master and the servant. In this land of bigotry the poor fellow would have been denied a grave, had I not, when he was dead, sent for a priest, who however on arriving turned unfeelingly to me, exclaiming " Umph ! he has died without confession —his soul is lost, and it will be needless for me to see him." Yet my having called in a priest was sufficient declaration that the deceased was a Catholic, and by the kind assistance of Mr. Dall I found no difficulty in having him interred in the Campo Santo with the ceremonies of the Catholic Church. On Sunday evening (June 4,) Mr. Dall and two other American gentlemen joined our little procession, carrying candles; and I saw my poor departed servant buried with proper decency.

June 5.—We visited this morning the public prison, filled with a half-naked ill-looking set of people mixed together in one common court, without any distinction on account of crime. Some were playing at cards,—others spinning thread to procure means for subsistence; the guiltless man and the murderer enjoying the same degree of liberty,—viz. the range of the large court. The place was tolerably clean, but possessing to us an appearance of misery, which would not be felt by its inmates, accustomed as the lower classes here are to sleep on the bare ground. There was neither table nor bench, and the only furniture I saw was a huge pair of untenanted stocks. We afterwards went to the Carmelite church under the escort of a Spanish merchant, who procured us admission to the gardens also. The building is on an immense and magnificent scale, remarkably clean and airy, and having spacious corridors, with cells capable of containing a far greater number of monks than at present occupy them. The church is crowded with ornaments, which in some places have a grand effect; and there is a portion, over the door of the principal chapel, of most intricate and handsome carving in stone, reaching to the roof of the church. In several other parts the fretwork is well executed, but with little regard to the rules of architecture.

Señor Tresguerra was at this time erecting a fine altar-piece of stone, which if not ruined by gilding cannot fail to be admired for its simplicity and elegance   There are many pictures of Elijah the patron saint of the Carmelites, who is accordingly introduced in a variety of fiery chariots, some few of which are well painted.   In one picture illustrating that portion of the First Book of Kings, xviii. 44, where "there ariseth a little cloud out of the sea, like a man's hand," the Virgin Mary in a splendid mantle is represented as sitting on the top of this cloud, with her hands closed in prayer; and a long account of the rain in the days of Ahab having been procured by her intercession, is subjoined in gold letters at one corner of the picture.   I could not resist remarking to our guide, that the Virgin in an Old Testament picture was rather misplaced: but he very gravely assured me that on this occasion she actually did appear, whatever the Bible might say

about the matter. In the choir and galleries of
the church are pictures of several well-authenti-
cated miracles, particularly one of a poor nun, who
being sick, had dared to provide a boiled fowl for
her supper, contrary to the rules of her order,
which forbade all use of flesh; she had just stuck
her fork into the prohibited morsel, when the Virgin
appeared to her,—like Sancho's physician at Ba-
rataria,—and prevented the unrighteous banquet,
enjoining a long list of penances to the poor sick
sinner for her crime. The Virgin, in fact, is introduced in Mexico into every painting, either before or since the flood, to the total exclusion of
that homage which is due to God alone.

The Carmelites are very rich, and possess immense estates and Haciendas, which according to
report will ere long be applied to the use of the
Government. Having examined the church, we
rambled through the extensive walled garden attached to it, round which were straight walks
shaded almost to darkness by luxuriant vines.
Apple- and pear-trees were very abundant, but

weeds more so. No taste was displayed in the laying-out the ground, which was irrigated from a large tank. I here saw the Grana, or Cochineal, in great abundance on some plants of Tuna, but it was never collected either for sale or use.

June 6.—We were so late in arranging our cargoes that we could not set out, as I had intended, for Zacatecas; and in the evening we rode to the Sanctuario de Nuestra Señora de Guadalupe, which is situated at a short distance from the town. An old lady, with an enormous goitre, introduced us to the church, where the Padre, her brother, sat praying with amazing volubility; and we walked about as we pleased, admiring the neatest temple we had seen, ornamented entirely in white and gold, and having few other pictures than those immediately relating to the Virgin of Guadalupe, whose legend is painted on the compartments round the dome. Her story has often been told by Mexican travellers, but I shall still give my version of it, after I shall have described my visit to her church near Mexico.

We were afterwards treated with cigars, chocolate, and politics, in which, as far as related to the disturbances in this country, the Padre showed himself well versed, having during the revolution headed a party of seventy Rancheros near Léon, who destroyed 450 regular Spanish infantry. I afterwards heard that our friend, at that time the Padre of Saltos de Barra, had been one of the most active of the many priests who took up arms against the Spaniards. Before leaving the Sanctuario he treated us with a view of San Luis from one of the high turrets, and waited patiently while I attempted a sketch of this most beautiful prospect. I found that Flores was still too ill to travel, and therefore left him behind, hiring another man to attend my horses.

June 7.—We quitted San Luis at 6 A.M. having sent forward the greater part of our luggage by Arrieros, and rode four or five miles over a plain thickly planted with maguey, from which the town is supplied with pulque. We met several of the natives with their donkeys laden with jars of this favourite liquor, and drank some sweet and recent from the plant. At about three miles from San Luis on the road to Zacatecas, the view on looking back upon the town is very beautiful. We soon exchanged the smooth road for a stony one, and ascended the Cerros of "San Miguel Miskitiki," over which the travelling was as rough as it was possible to be. The rocks were of porphyry, chiefly of a red colour, and coated abundantly by beautiful mammellated chalcedony, which with fragments of jasper lay in great quantities in the road. In five leagues we descended a gorge in the mountains, and came suddenly on the retired romantic little village of San Miguel Miskitiki, situated in a deep rocky glen and half hidden by its flourishing trees. A mountain stream of very clear water runs through the ravine, and is crossed by a small picturesque bridge of two arches. Leaving San Miguel we came to higher ground with a flatter and better road, and at a quarter after one arrived at the Hacienda "La Parada," ten leagues from San Luis. The land is here extensively cultivated, and in many places requires to be watered by irrigation from a large tank, in which I found several wild ducks and a very small species of diver. There is a kind of village and a church here, and we obtained lodging in a Meson, which according to custom in this country afforded nothing but shelter, water, and one small candle which is always given to the guests without extra charge.

June 8.—Thermometer 57°. At seven we started over a pretty good road, but a parched-up dusty country, thinly clothed with mimosa bushes, the maguey, and tuna. I here first saw a humming-bird, which flew off before I could approach it. Hares and partridges were very abundant, and I killed some for our dinner. On the road we met a coach drawn by eight mules, and having eight others running before it, guarded by a dozen odd-looking soldiers. Before noon we passed a small Rancho crowded with fine cattle, where we stopped for a time at a hut, and found a most excellent breakfast of fried pork and onions, milk and tortillas, all which had been prepared for a Ranchero. We left the poor man nothing but the empty dishes, and the reals we had paid for his breakfast. At one, after having passed over a plain thickly covered with maguey, we came to an establishment for making Vino Mescal. This is an ardent spirit distilled from the heart of the maguey, which being deprived of its leaves to the very bottom of the root is well bruised and boiled; it is then placed in immense leather bags suspended from four large stakes, and allowed to ferment, with the addition of pulque and the branches of a shrub called " Yerba Timba" to assist the fermentation. These leather bags each hold about two tuns. The liquor when sufficiently prepared is baled from them to the still, which is inclosed in an immense staved and hooped vessel, like an overgrown cask, from whence the distilled spirit flows by a spout made of a maguey leaf. This stands over a subterraneous fire, and the cooling water is contained in a large copper basin, which fits into the top of the cask and is removable at pleasure. The " Vino Mes-

cal" is then stored in entire bullock's hides, of which
we saw a room quite full, and their appearance resembled a quantity of cattle hung up by the hocks
and deprived of the feet, head, and hair. Each of
these would, I conceive, hold about two pipes, and
they had a spout in the lower end formed from the
skin of the neck. When struck by the open hand
they gave forth a jingling sound as if containing
glasses. The Vino Mescal is sent to the market in
goat-skins, of which we saw some milk-white ones
lying inflated in readiness; and not far from the
house was a flock of many thousand goats of the purest
white, with a quantity of brood mares and fine cattle.
The conductor and some of the other people gave
evidence of the strength of the commodity which
they manufactured, for they were half-drunk, very
smoky, and extremely dirty. But we were surprised
in the midst of this scene by the apparition, from a
little mud hut where the men resided, of a young
girl of about thirteen, perfect in beauty and figure,
and as fair as a European. At four we reached the
Hacienda of El Espiritu Santo (The Holy Ghost).

These are startling appellations to an English ear,
but in this most Catholic country they are quite common. At Tampico there is "The shop of the Deity;"
in San Luis, "The Holy Trinity;" while "The
Divinity," "Jesus Christ," and other sacred titles,
are designations for drinking-houses and the mean-
est shops. The fields at the Hacienda were in the
highest state of cultivation, surrounded by good
walls; and here we saw the first field of wheat,
quite ripe for the sickle. There are large ponds
dammed up in the valley by "Presas," or walls,
of excellent masonry, well plastered and white-
washed; and the village of the Hacienda displayed a neatness and air of good management rarely
to be seen in Mexico. The Hacienda abounds in
fine cattle, horses, and mules. In the lagoon we
saw numbers of ducks, and a white heron. The
prosperity of this place is attributed to the owner
having armed his people in defence of his property
during the devastating revolutionary war; and its
contrast with some Ranchos which we had passed
on our day's ride was very striking. There we

saw the houses roofless and in ruins blackened by fire, and had ridden over plains still bearing faint traces of the plough; but the Rancheros who had tilled the ground had been murdered with their whole families during the war. In the space o' forty miles we passed no fewer than fifteen crosses set up at the road-side, to mark the spot where an assassination had been committed, and to claim the traveller's prayers for the soul of the unfortunate victim. This part of the road was still considered unsafe; we therefore took the precaution not to separate from each other, and carried our arms in readiness. We lodged in a very good Meson, having its rooms on two sides of a large court, in which we found a convoy of 130,000 dollars on their way to the coast, guarded by half-a-dozen ill-looking ragged soldiers, with ineffective muskets, and horses which could scarcely support themselves *.

June 9.—We set out at seven over a dusty but excellent road, through the same barren unin-

* Our journey had been fourteen leagues.

VOL. I.

teresting country, destitute of water, and bounded by distant mountains. Here the whirlwinds were very numerous and troublesome. In the course of our ride we saw several wild deer and some very small humming-birds. On this day we passed the cairns of three murdered persons; and having ridden ten leagues, arrived soon after noon at the Government salt-works and Hacienda of Las Salinas, where we obtained a room in the Meson of " Jesus of Nazareth." The Hacienda contains about five hundred persons, and is situated on an arid plain, near which are the marshes whence salt in an impure state is procured. This is consumed in great quantities at the " Haciendas de Plata *" of the mining establishments, where it is used in the process of amalgamation. There is a military establishment and barrack here, inclosed within extensive walls pierced with loop-holes for musketry, and surrounded by a ditch, of about ten feet in depth and as many in width, cut in the solid lime-

* The place where the ores are reduced, and the pure metal extracted.

stone. A hundred Royalist soldiers garrisoned this place during the wars; and it was never once taken by the insurgents, although there was much fight-ing on the plain. The houses here are built of sun-dried bricks, of a much whiter and better qua-lity than any I had before seen.

Saturday (*June* 10).—Leaving Las Salinas at se-ven, we rode over the swamp, which was covered by a brittle crust, and examined some heaps of the salt collected for sale. They were of the second gather-ing, called " Saltierra," and so blended with earth that their quality was only perceptible by the taste. This is sold at four reals the Fanega* (bushel), but "La Primera Flor" (the first efflorescence, or "Sal Blanca") is from three to three and a half dollars the fanega. It is conveyed to the Hacienda at the Veta Grande in cars drawn by bullocks, generally at the rate of three reals the fanega. As the Sa-lina extends three miles in length and one and a half in width, it is natural to suppose that the

* This is the same measure as is used in Spain, and is equal to 1·599 English bushels.

Government derive a considerable revenue from it,—the Hacienda of the Veta Grande of Zacatecas alone consuming in Saltierra and Sal Blanca to the amount of 35,000 bushels per annum, even in the present comparatively unproductive times. Our whole journey on this day was over an excellent, though dusty road, through a desert only enlivened by the numerous spiral whirlwinds which half-buried us at times beneath the cloud they created. At the distance of five miles we came to a cluster of the most miserable huts I ever met with, called El Pozo; few of them had any side walls, and merely consisted of a low sloping roof just high enough in the centre to allow of a person's standing upright. It is impossible to conceive how the inhabitants support themselves for some part of the year; all is desert around them, and the salinas only afford a few months employment. At three leagues from our outset we passed about twelve miles to the northward of a long Sierra, where the mining Real de los Angeles is situated, and on its left, at about thirty miles, is a sugarloaf Sierra de Altamira just discernible, on which is the Real de los Asientos de Ybarra, from whence the city of Aguas Calientes is only eight leagues. Still further on its left again, we saw a rugged white Sierra, containing the Reales of Chipinqui and de los Milagros. Riding about two leagues further, our people pointed out on the right hand the Real and church of Ramos; and at last on rising a small hill we saw the Cerro of Zacatecas, about forty miles to the westward. We now considered ourselves as having fairly entered on the mining district of Mexico; and a more desolate dreary country than this appeared in the month of June, scarcely exists on the face of the globe,—after excepting the great Desert in Africa, and the Polar Regions. We completed our day by arriving at a wretched mud village named La Blanca, and put up at a ruined Hacienda de Plata, having travelled twelve leagues, in which we saw five cairns and crosses. At about four miles before reaching this wretched place, we passed La Laguna y Rancho del Moro, lying at a little distance to our left. Considerable quantities of salt covered the ground at this place, which a number of people were scraping up and putting into a bullock cart. The flat valley was covered by a weak kind of parched grass, on which above three hundred brood mares with their colts and a large flock of sheep were feeding. Here I saw for the first time a Coyote, or jackal, at which I had an ineffectual shot; and we also observed that the ground squirrels* were very numerous.

June 11.—Leaving La Blanca at eight, we proceeded over a country beyond all description gloomy and barren. At five leagues we passed a cluster of hovels called Franosa, a little on our left; and in five leagues more began a very gradual descent into a sterile valley of five or six miles in width, which brought us to the foot of the Cerro of Veta Grande. A gentleman from the Veta met me, and we rode up the rugged ravine amongst the mountains to the mine and village of **La Veta Grande**, which are situated nearly at its head, about

* " Ardilla"—" Tusa."

five miles from the plain, or rather from the Hacienda de la Purissima Concepcion de Sauceda, where the ores from the mine are amalgamated. I had prepared myself to be as much disappointed with this place, after the description given me, as with almost every thing else I had seen since landing in the much vaunted Republic of Mexico; but found it ten times worse even than I could have anticipated. High steep hills covered with stones and the parched remains of last year's miserable allowance of grass, surrounded the Veta. Not a tree, and scarcely any thing which by a stretch of imagination might be conjured into a bush, was to be seen; and often have I travelled over a more productive and verdant soil, in the Polar Regions, than is to be found within some miles of the Veta Grande. The people, who are either employed in or connected with the mines, live in a little village along the side of a small Barranca; and the "Casa Grande," into which I was ushered, was filled at this time with a variety of people of every description,—all messing at one table, and sleeping two or three in a room. It was still my fate never to have a whole apartment to myself, and want of accommodations again obliged me to have a fellow-lodger. I had consolation however for various discomforts and inconveniences, in finding Dr. Coulter, who was in charge of the Veta Grande until my arrival, a most agreeable, scientific, and gentleman-like man.

CHAPTER III.

Village of Veta Grande—Feast-day—Mine of San Bernabe—
Zacatecas—Rains—Dislike of the Natives to the English—
Village of the Sauceda—Colegio of Our Lady of Guada-
lupe—Christening—Ruins of a large Indian City now called
Los Edificios—Ancient Causeways connected with it—
Mal Paso—Chase of the Coyote or Jackal.

Our residence at Veta Grande certainly possessed one charm—in the magnificence of its view.
From our windows on clear days, and the air was
generally very pure and transparent, we could
look over the vast arid plains in the north to the
remarkable mountains of Catorce, distant about
two hundred miles: yet even this extensive *coup-
d'œil*, when the novelty had ceased to charm,
was but increasing the feeling of desolation; the
eye in all that extent wandering over barren
plains, scarcely enlivened by a single tree, and
wholly destitute of water. To compensate for the
dismal exterior of the Veta Grande, the mines

were promising well, and varying in their produce from five to ten thousand dollars weekly. Every process was still carried forward on the Mexican system, which with a few slight and gradual alterations was to be continued, after some abuses should be corrected, and the useless members weeded out of the community.

June 18.—On our first feast-day the village of Veta Grande appeared to have undergone some magical change, and to be peopled by a different race from those who had figured during the week. Fine shawls, brilliant-coloured gowns, silk stockings and white satin shoes, were flashing like so many meteors amongst the mud huts; and in the evening I accepted an invitation to go to an exhibition of Maroméros, or rope-dancers, in company with two maiden ladies, sisters of a certain Don Jesus, who kept a little shop, and was one of the principal gentlemen in the town. It was a fine moonlight night, and we walked to a small mud amphitheatre usually appropriated to cock-fights, where we found the tight rope stretched, and a numerous party-coloured audience assembled. The theatre was open to the clear starry sky, and illuminated by four flaming piles of the Ocote or candle-wood, placed in iron cradles on the summits of tall poles. The whole scene was very novel and striking to me, as the miners and villagers lay extended and lounging on the earthen seats wrapped in their variously striped serapes; while five of the "milicia" moved about in the crowd to preserve order.

The ladies kept us plentifully supplied with cigars, which they also smoked abundantly; and in our turn we purchased sugar-plums and sweet cakes for them during the very short intervals of smoking.

The rope-dancing was tolerable, particularly by a very fat old woman gorgeously attired, who seemed in a terrible fright lest she should have a fall. A boy of about twelve years of age quite astonished us by his activity and the variety of his postures and contortions, which far exceeded any thing of the kind I had ever seen in Europe. The tumblers were attended by a clown, who with a blackened face and much talking greatly delighted the company. The performances were closed by a " Comedia" in front of a ragged sheet.

June 21.—Captain Vetch, first-commissioner of the affairs of the Real del Monte and Bolaños Companies, arrived from the latter place, for the purpose of closing accounts with the lessees of the mines, and putting affairs in order.

June 22.—An accident which happened this morning in the mine of Gajuelos gives a good example of the force of prejudice amongst the natives. One of the miners, in descending by the very awkward ladder way, missed his footing, and in his fall carried another poor wretch to the bottom with him. He who first fell was instantaneously killed, and the other man died in two days after the accident. I went with Dr. Coulter to see the living sufferer drawn up the shaft in a net; but we were not permitted to examine his hurts, until the priest, who also attended, had confessed him; it being a law in this strange country, that if a man is found even stabbed and bleeding in the street, no good Samaritan may venture to stop his wounds until the Alcalde has seen him and the Padre has taken his confession. Thus the law and the gospel, instead of saving, are often the means of destroying the life of an unfortunate being, who by timely assistance might have been snatched from death.

June 25.—The periodical rains, which had been anxiously expected for some time but had been kept back by a continuance of easterly winds, now set in with considerable force.

June 28.—On this day we rode into Zacatecas, which was my first visit there. The road to it lies over the summits of the high ridge of mountains, and is about as uninteresting a route as can well be imagined. Here and there the buildings and Malacates of various mines break to a certain extent the monotony of the prospect; and in a hollow to the eastward of the road lies the remarkable mine of San Bérnabe, still in work, and reputed to have been one of the first which was opened after the Conquest. The individual who then possessed it soon acquired great wealth; and although of obscure birth, married the daughter of the Viceroy, which gave rise to the following lines, since become a kind of standard saying amongst the miners of the neighbourhood:—

> *" Si la Mina San Bérnabe*
> *No daria tan buena Ley*
> *No casaria Pedro Barra*
> *Con la Hija del Virey."*

The first appearance of Zacatecas as the traveller approaches from the northward is peculiar and pleasing. The city stands in a deep basin at the foot of a picturesque and abrupt mountain, called the Bufa; and the entrance by the suburbs is through a gravelly water-course, in which groups of women are seen washing clothes. We paid a visit of ceremony to his excellency general Lobato, some short time since a very respectable cobbler at Jalapa, and now commander in chief of the " Free and Sovereign State * of Zacatecas." He was un-

* Estado libre y soberano.

well and confined to his room; but we were received by his lady, a thin talkative little woman,
who abused both miners and mining in most un-
qualified terms; and by her sister, a large, greasy,
half-dressed maiden, with black moustachios and
nut-brown teeth. The ladies sat huddled up in a
corner smoking; and the tiled floor, on which reposed an immense dog and her puppies, was
strewed with extinguished cigars and their ashes,
cabbage and lettuce leaves, and other filth which
had fallen from five bird-cages hung along the
centre of the room.

Two unshaven and unwashed cavaliers were
paying their morning compliments to La Generala,
and the whole scene was such, that I retired from
it with no very favourable ideas of the *beau monde*
at Zacatecas. Having made equally gratifying
visits to one or two other of the most distinguished families, we rode home in the rain, which now
fell regularly every day at about two or three
o'clock in the afternoon.

From our great elevation above the extensive
plains we were enabled to command most mag-

nificent views of the progress and formation of the clouds as they swept in large black masses, at an equal height, over the low grounds, and on reaching the mountains bounded as it were from summit to summit accumulating in their progress all the smaller vapours, and then with heavy thunder and vivid lightning breaking, and deluging for a few hours the whole country beneath. The outline of the storm was invariably so clearly defined, as it rushed towards us from the east, that while one half the heavens was blackened by the growing tempest, the western sky was of a bright and cloudless blue, and the plains glowing beneath the dazzling rays of the sun.

July 9.—On the 9th a party of English artificers and miners, under the charge of my friend Mr. Tindal, arrived from Real del Monte, and passed through Zacatecas at the time it was most crowded with people, who on Sundays flock from the neighbourhood to attend the market. On these occasions they generally get drunk, when they become quarelsome and too frequently use their knives against each other. It was an unlucky moment for strangers to appear amongst them, and they availed themselves of it to quarrel with the English and to throw stones at them;—had not a party of the city " milicia" been sent to protect the new comers on their way to the Veta Grande, some serious consequences might have ensued. The Custom-house officers having taken it into their half-tipsy heads that the baggage of the travellers contained some arms, stopped it all in the middle of the town, and Mr. Tindal and I were obliged to ride there to settle matters. By humouring the crowd, who were already ripe for mischief, we kept them in tolerably good temper; but no sooner were our backs turned, than we were saluted with a half merry half saucy hiss, and they honoured our retreat with a few stones.

Considerable ill-will was also manifested towards the strangers by the miners at the Veta; and when they appeared singly, they were pelted. An attack was actually made at night on the door of the house in which they were quartered, and it was battered with stones.   Four ringleaders of the assailants were taken up and imprisoned, and on the following morning a threatening paper which had been pasted on our stable gates and on the door of the Alcalde was brought to us.

July 10.—Captain Vetch addressed a very strong remonstrance to the authorities of the village, respecting the conduct of the natives; but on the succeeding morning similarly intolerant papers were again placarded in a kind of doggrel rhyme.

On the 12th I went to reside at the Hacienda de Sauceda, at which the silver is beneficiated *. This establishment is situated about five miles N. E. of the Veta Grande, at the beginning of the ascent from the plains.   It was decided that this retired spot should be my head-quarters; and having discharged the native Administrador, who defrauded us of a considerable sum of money, I assumed his office and conducted the concern.

The little quiet village of the Sauceda, with the comparatively gentle manners of its inhabitants,

* Extracted from the ore.

were quite delightful after the bustle of the crowded Veta; and it was not one of my smallest gratifications, to enjoy for a whole week a room to myself. The eastern front of the gallery in which I lived looked into the " Patio" or large paved court, in which the silver ores are submitted to the process of amalgamation; and beyond its walls the eye could roam for twenty-five miles over a plain terminated by a low line of mountains, at whose foot a white church pointed out the situation of the Real or mining district of Ramos. The recent rains had achieved wonders in the rugged barrancas; many places having changed their brick-coloured hue for a lively green, while small bushes peeped here and there from the crevices of the rocks near the water-courses; the plains too looked fresher and more cheerful;—in fact, I had changed my residence from one of the most desolate spots which could be imagined, to a better situation, and all nature wore a more pleasing aspect in consequence.

Of our domestic affairs I may mention that the

first English child was born at the Veta Grande; and at the same time a little humming-bird died, which I had kept for nearly a month on sugar and water slightly impregnated with saffron. It greedily sucked this mixture from a small quill; and I am sure that with constant attention these little creatures might be kept for a long time.

July 15.—The first-commissioner left for Real del Monte, and a party of us accompanied him to the Colegio of Our Lady of Guadalupe, to which we had been invited. This convent is situated at the foot of the mountains about a league to the eastward of Zacatecas, and surrounded by a little village which has arisen on the sanctified spot. It has a very pleasing appearance, being embellished by about two dozen trees. The Fathers received us most kindly; and with the attentions of the Padre Guardián and Padre Maldono (Macdonald, an Irishman), we passed a very agreeable evening; were regaled with a good supper, and a separate cell with a clean bed was provided for each of us.

July 16.—Making an early tour of the convent,

we first visited its library, consisting chiefly of religious works bound in parchment, to the amount of 11,000 volumes. None of them were particularly remarkable either for type or age, and I inquired in vain for old Mexican MSS. or objects of antiquity; which, as this society of friars are all Missionaries, I expected they would have collected on their visits to the more remote districts and tribes in New Spain. The Colegio, which is large, is profusely ornamented with very ill-executed paintings, chiefly relating to the life of San Francisco, who in power and miracles very far exceeded the Saviour, the latter being actually represented attending him as a menial servant.

One picture particularly amused me, as the best specimen of the Fuseli school I ever saw. It represents the Jewish council debating upon the proposed seizure of our Saviour. They are a grave and venerable party, but each has, perched either on his head or shoulders, a devil, who is whispering his wicked thoughts.

All these imps, however, are painted with the

most laughably roguish snouts and eyes, and the
oddest claws and tails imaginable; while the el-
ders, perfectly unconscious of their strange as-
sociates, are very serious, communing with each
other. The church and chapels have nothing remarkable, except one very highly esteemed show,
where Joseph and Mary in gorgeous apparel are
kneeling near a wilderness of gold tinsel wire;
while around them are a confused variety of little
images not a twentieth part of their size. Amongst
the multitude is one female Chinese figure, with
the usual dead white face and long eyes, and another Chinese woman bearing a child made of soap-
stone. The most grotesque, however, is a little
drunken Dutch farmer in leather breeches and a
red waistcoat, who is placed very properly in the
foreground to prevent the scandal his company
would throw on the other idols. One eye is open,
and its fellow is closed, with an air of slyness and
roguery which gives a most comical expression to
his tipsy face. This is perhaps the first Dutch
saint which has ever been worshiped in Mexico.

Nothing could be more refreshing than the extensive gardens, crowded with apple-, pear-, fig-, and quince-trees. There were also some vines, pomegranates, peaches, and apricots. Immense rose-hedges rendered the walks in many places shady and sweet, and the whole was well cultivated, —a kind of Oasis in fact, in the desert of Zacatecas. We were all much pleased with our visit, which gave great satisfaction to the Fathers, who were happy to entertain any one from the mines, whence they draw the chief part of their very uncertain revenue in return for confessions and masses, as well as by begging, according to the law of their order, from all the cottagers.

The poor friars of Guadalupe are a much-enduring race, and should not, I conceive, be classed with the herd of drones who batten so uselessly on the public in Mexico. These do actually endure all the poverty which their vow enjoins, and their whole life is devoted to voluntary suffering. They have no personal property beyond one coarse gray woollen dress, which is never changed until worn into holes; and it then, having fully attained the odour of sanctity, sells for twenty or thirty dollars, as the burial garment of some devotee, who is supposed to smuggle himself into heaven in so holy an envelope. They wear neither shirt, stockings, nor any other articles of dress than a pair of sandals; and in the high cold mountains of the northern states, or low in the burning plains of the Tierra Caliente, no change of habit appropriate to the climate, is permitted by the rules of the order.

The Colegio de Guadalupe was founded expressly for the purpose of furnishing missionaries " para conquistar," or for the conversion of the Indians of Texas, California, or indeed any barbarous northern tribes; and about half the friars of the establishment are constantly absent on these pious errands. A vast number of these poor men have perished from absolute want on their weary journeys, being sent forth without money or even an animal to carry them, and dependent on charity for their subsistence. Many also have

been barbarously sacrificed by the wild Indians:—
yet the missions are constantly maintained, and
with success; since in the most retired and inhos-
pitable parts of the northern states, communities
of some thousand Indians may be found living
under the spiritual guidance of one or more of
these poor friars; although the faith which is
taught them is so modified from the Mexican religion, in order to suit their habits and capacities,
that the name of "Christiano" is all of Christianity
which they learn.   Amongst a barbarous people
the Missionaries, although themselves most grossly
ignorant, appear gifted with almost supernatural
acquirements, scarcely inferior to the Saints and
Martyrs whose absurd legends they so strenuously
inculcate.   The Padres, who reside in their turns
in the Colegio, undergo a life of perpetual mortifi-
cation, with but little rest;—constant prayers with
meagre diet. In addition to this, they retire at
seven in the evening to a darkened room having
only one small taper burning before a crucifix:
here for one hour they flog themselves with a

small whip of twisted wires, called a *disciplina,* and sing the " Miserere" during the penance. I have in my possession one of these instruments of tc.ture; and were every penitent to give himself but half a dozen good strokes with it, I should believe their assertion, " that the hall of contrition is sometimes covered with blood." A few of these poor enthusiasts may occasionally strike a good fair blow; but human nature is such, that I conceive honest Sancho Panza's hint for unseen flagellation is not thrown away upon the greater part of them.

After having undergone the discipline, at which it is said some of the poor wretches faint, the whole community, with downcast eyes, arms folded on their breasts, and in perfect silence, retire to their respective cells, where they may sleep or meditate until aroused at midnight to perform their religious exercises for the space of three hours. Novices are admitted to the order at sixteen years of age; at seventeen they may profess. Before the expiration of the first year, however,

they are permitted to retire; an advantage of which
they rarely avail themselves, the convent being so
much revered in the neighbourhood that an admission to it is thought as honourable as retirement
would be considered disgraceful.   At twenty-four
they may perform the sacrifice of the mass; and at
twenty-six hear confessions.   The Padre guardián
is elected every three years; but his situation, far
from being a desirable one, entails on him more
labour and trouble than is imposed on his brethren.
The poor recluses have but two enjoyments permitted to them. On Thursdays and Sundays they
play at quoits for three hours, and on great festivals no discipline is required.

We returned to the Hacienda by the plains, now
covered with bright young herbage, which appears
with the rains and withers when they cease.   A
cloudless sky and glowing sun had called from
their hiding-places innumerable little animals of
the marmot tribe, called " Tusa." This pretty
creature is somewhat smaller than an ermine, and
is of a yellowish gray colour; it burrows exten-

sively in the plains, and where large communities of them exist, the footing is dangerous for horses.

July 23.—This was a day of much bustle and confusion, in which we had a full opportunity of ascertaining the extent of the religious prejudices against us. The infant of one of our artificers, to whom I was to be godfather, was to be christened in the church of the Veta Grande; and as several English children had received that ceremony without opposition or comment in the city of Mexico, no impediments were expected here; the two ceremonies of the Catholic and Protestant churches being, with the exception of language, nearly the same. The day was passed in long letters and objections, which ended in an injunction that the heretical godfather was not to approach the baptismal font. It was late at night before all was settled, and our party proceeded to church; but as the other English were not permitted to be present at the ceremony, I of course retired with them; and the child's father, with a native servant whose knowledge of English condemned him also as a heretic, were turned out with the rest. I walked indignantly home, and was soon followed by those who had waited in the sacristy, bringing the baby, which after all was not christened by the name that was intended, but by some fancy of the very reverend and most Christian Padre was called José Bonaventura, after which the intended name Jorge (George) was added.

My retreat very much discomposed the priest and his attendants, who imagined they should make a very good harvest of poor little José Bonaventura's christening. And while all the business was pending, and I was waiting in no very good humour the result of a discussion with which I was tired, the following most agreeable hints were supplied to me, as the intended godfather.

1st. The church would be splendidly illuminated in honour of the English, for which of course the Padre would expect an extra fee . . . . 12 *Dollars.*

2ndly. The organist intended doing himself the honour of playing an anthem after the ceremony 4

3rdly. The sacristan on so joyful an occasion could not possibly be presented with
a smaller sum than . . . .   *Dollars.*   4

4thly. A notary, from a disinterested wish
to render the ceremony as respectable as
possible, would have the pleasure of being present to register it 4

5thly. Two little choristers would put on
their red cloaks in honour of the event . 4

6thly. A scramble of medios (silver three-
pences) were to be thrown by the delighted Padrino amongst the joyous crowd . 10

7thly. Medios spick-and-span new were, according to an old custom, to be presented to every acquaintance and decent person in the crowd* 10
                                         Total . 48

To say nothing of the gifts which I was to make to
the native godmothers; one of whom, the fat old
housekeeper, rustling in black silks and smoking

* This gift is also usual at weddings, and is called "Bola."

like a furnace, I did make happy by the present of a Birmingham comb in the form of a tiara.

Had the young Christian been the child of a prince he could not have created a greater sensation; but as both father and godfather had been expelled the Catholic temple, no one was present to answer the demands after the baptism. The organ gave forth no joyous peal; the notary did not make his appearance; and although the sacristan and choristers remained in their full uniform, no one was found to reward them for their attention. At last, to the total discomfiture and astonishment of the Padre, the discovery was made that the babe was the son of a stone-mason, and that he would be paid accordingly !

July 29.—An intelligent Cura having given Mr. Tindal and myself some information respecting the ruins of an ancient Indian city about fourteen leagues to the southward of Zacatecas, on the road to Villa Nueva, we took advantage of a holiday and rode out to see them. Passing through Zacatecas, we travelled four hours in the first instance over mountains, and latterly on an extensive plain devoted to pasturage, until we reached a small picturesque Rancho, called El Fuerte. Beyond this, at about one mile, we passed a massive and magnificent Presa (or dam), made to form a large tank of water. It extended about four hundred yards across a gentle valley, and was of excellent masonry, supported by immense buttresses, so as to resist the weight of water at its back, and which was a store for the supply of the beautiful Hacienda of Mal Paso, stretching some leagues to the southward. The plain we had ridden over was destitute of trees, but in the vale of the Hacienda they abounded, and were in rich foliage. Large herds of fat cattle and horses were grazing in the extensive pasture districts: all bespoke wealth and good management, and was the more striking to us after having lived amidst the barren hills of the Veta Grande. The country about here was thickly strewed with masses of gray porphyry, and chalcedony in nodules was very abundant.

Having ridden five leagues beyond Mal Paso,

we arrived at sunset at the N.W. corner of the isolated hill on which the ruins we were in search of
were situated; and having examined a small artificial cave with two entrances scooped in the porphyritic rock, we rode on another league to the
Hacienda La Quemada, where with great difficulty
and much entreaty we procured a night's lodging
in a hut.

July 30.—On the following morning we set out
on our expedition to the Cerro de los Edificios,
under the guidance of an old Ranchero, and soon
arrived at the foot of the abrupt and steep rock on
which the buildings are situated.   Here we perceived two ruined heaps of stones flanking the
entrance to a causeway ninety-three feet broad,
commencing at about four hundred yards from
the cliff.

A space of about six acres has been inclosed by a
broad wall, of which the foundations are still visible,
running first to the south and afterwards to the
east. Off its south-western angle stands a high mass
of stones which flanks the causeway.   In outward

appearance it is of a pyramidal form, owing to the quantities of stones piled against it either by design or by its own ruin; but on closer examination its figure could be traced by the remains of solid walls to have been a square of thirty-one feet by the same height: the heap immediately opposite is lower and more scattered, but in all probability formerly resembled it. Hence the grand causeway runs to the N. E. until reaching the ascent to the cliff, which, as I have already observed, is about four hundred yards distant. Here again are found two masses of ruins, in which may be traced the same construction as that before described; and it is not improbable that these two towers guarded the inner entrance to the citadel. In the centre of the causeway, which is raised about a foot and has its rough paving uninjured, is a large heap of stones, as if the remains of some altar, round which we could trace, notwithstanding the accumulation of earth and vegetation, a paved border of flat slabs arranged in the figure of a six-rayed star.

We did not enter the city by the principal road,

but led our horses with some difficulty up the steep mass formed by the ruins of a defensive wall, inclosing a quadrangle 240 feet by 200, which to the east is still sheltered by a strong wall of unhewn stones eight feet in thickness and eighteen in height. A raised terrace of twenty feet in width passes round the northern and eastern sides of this space, and on its S. E. corner is yet standing a round pillar of rough stones, of the same height as the wall, and nineteen feet in circumference.

There appear to have been five other pillars on the east, and four on the northern terrace; and as the view of the plain which lies to the south and west is hence very extensive, I am inclined to believe that the square has always been open in these directions. Adjoining to this we entered by the eastern side to another quadrangle, entirely surrounded by perfect walls of the same height and thickness as the former one, and measuring 154 feet by 137. In this were yet standing fourteen very well constructed pillars, of equal dimensions with that in the adjoining inclosure, and arranged four in length

and three in breadth of the quadrangle, from which
on every side they separated a space of twenty-three
feet in width, probably the pavement of a portico of
which they once supported the roof. In their construction, as well as that of all the walls which we
saw, a common clay having straw mixed with it has
been used, and is yet visible in those places which
are sheltered from the rains. Rich grass was growing in the spacious court where Aztec monarchs
may once have feasted; and our cattle were so delighted with it that we left them to graze while we
walked about three hundred yards to the northward, over a very wide parapet, and reached a perfect, square, flat-topped pyramid of large unhewn
stones. It was standing unattached to any other
buildings, at the foot of the eastern brow of the
mountain, which rises abruptly behind it. On the
eastern face is a platform of twenty-eight feet in
width, faced by a parapet wall of fifteen feet, and
from the base of this extends a second platform
with a parapet like the former, and 118 feet wide.
These form the outer defensive boundary of the

mountain, which from its figure has materially favoured their construction. There is every reason to believe that this eastern face must have been of great importance. A slightly raised and paved causeway of about twenty-five feet descends across the valley in the direction of the rising sun, and being continued on the opposite side of a stream which flows through it, can be traced up the mountains at two miles distance, until it terminates at the base of an immense stone edifice, which probably may also have been a pyramid. Although a stream (Rio del Partido) runs meandering through the plain from the northward, about midway between the two elevated buildings, I can scarcely imagine that the causeway should have been formed for the purpose of bringing water to the city, which is far more easy of access in many other directions much nearer to the river, but must have been constructed for important purposes between the two places in question; and it is not improbable, that it once formed the street between the frail huts of the poorer inhabitants. The base of the large pyramid measured fifty feet, and I ascertained by ascending with a line that its height was precisely the same. Its flat top was covered with earth and a little vegetation; and our guide asserted, although he knew not whence he received the information, that it was once surmounted by a statue. Off the S. E. corner of this building and at about fifteen yards distant, is to be seen the edge of a circle of stones eight feet in diameter, inclosing, as far as we could judge on scraping away the soil, a bowl-shaped pit, in which the action of fire was still plainly observable; and the earth, from which we picked some broken pieces of pottery, was evidently darkened by an admixture of soot or ashes. At the distance of one hundred yards S. W. of the large pyramid is a small one, much injured and twelve feet square. This is situated on somewhat higher ground, in the steep part of the ascent to the mountain's brow. On its eastern face, which is towards the declivity, the height is eighteen feet; and apparently there have been steps by which to descend to a quadrangular space, having a broad terrace round it, and extending east one hundred feet by a width of fifty. In the centre of this inclosure is another bowl-shaped pit, somewhat wider than the first. Hence we began our ascent to the upper works, over a well buttressed yet ruined wall, built to a certain extent, so as to derive advantage from the natural abruptness of the rock. Its height on the steepest side is twenty-one feet, and the width on the summit, which is level with an extensive platform, is the same. This is a double wall, one of ten feet having been first constructed and then covered with a very smooth kind of cement, after which the second has been built against it. The platform which faces to the south, and may to a certain extent be considered as a ledge from the cliff, is eighty-nine feet by seventy-two; and on its northern centre stand the ruins of a square building, having within it an open space of ten feet by eight, and of the same depth. In the middle of the quadrangle is to be seen a mound of stones eight feet in height. A little further on we entered by a broad opening between two very perfect and massive walls, to a square of 150 feet. This space was surrounded on the south, east and west, by an elevated terrace of three feet by twelve in breadth, having in the centre of each side, steps, by which to descend to the square. Each terrace was backed by a wall of twenty feet by eight or nine. From the south are two broad entrances, and on the east is one of thirty feet, communicating with a perfect inclosed square of 200 feet, while on the west is one small opening, leading to an artificial cave or dungeon, of which I shall presently speak.

To the north, the square is bounded by the steep mountain; and in the centre of that side stands a pyramid with seven ledges or stages, which in many places are quite perfect. It is flat-topped, has four sides, and measures at the base thirty-eight by thirty-five feet, while in height it is nineteen. Immediately behind this, and on all that portion of the hill which presents itself to the square, are numerous tiers of seats, either broken in the rock or built of rough stones. In the centre of the square, and due south of the pyramid, is a small four-sided building, seven feet by five in height. The summit is imperfect, but it has unquestionably been an altar; and from the whole character of the space in which it stands, the peculiar form of the pyramid, the surrounding terrace, and the seats or steps on the mountain, there can be little doubt that this has been the grand Hall of Sacrifice or Assembly, or perhaps both. Amongst the stones of the altar Mr. Tindal killed a blue-tailed lizard, which is rare, and one of the most beautiful little creatures of the species, and the bright ultramarine blue of the tail is not excelled by any artificial colour. In a curious old book * I found a description of this creature, with the account of a superstition which exactly corresponds with one existing in Africa with respect to the Ourral, a large kind of lizard.—" They are poysonous, and thirst after the blood of breeding women; and they report that if a woman, or but

* Lionel Wafer's Voyage to the Isthmus of America. A.D. 1677.

her clothes do touch this creature, she will afterwards prove barren."

Passing to the westward, we next saw some narrow inclosed spaces, apparently portions of an aqueduct leading from some tanks on the summit of the mountain, and then were shown the mouth of the cave, or subterraneous passage, of which so many superstitious stories are yet told and believed. One of the principal objects of our expedition had been to enter this mysterious place, which none of the natives had ever ventured to do, and we came provided with torches for the purpose: unfortunately, however, the mouth had very recently fallen in, and we could merely see that it was a narrow well-built entrance, bearing in many places the remains of good smooth plastering. A large beam of cedar once supported the roof, but its removal by the country-people had caused the dilapidation which we now observed. Mr. Tindal, in knocking out some pieces of regularly burnt brick soon brought a ruin upon his head, but escaped without injury; and his accident caused a

thick cloud of yellow dust to fall, which on issuing
from the cave assumed a bright appearance under
the full glare of the sun;—an effect not lost upon
the natives, who became more than ever persuaded that an immense treasure lay hidden in this
mysterious place. The general opinion of those
who remember when the excavation was clear is,
that it is very deep, and from many circumstances
there is a probability of its having been a place of
confinement for victims. Its vicinity to the great
hall in which there can be little doubt that the
sanguinary rites of the Mexicans were once held,
is one argument in favour of this supposition; but
there is another equally forcible,—its immediate
proximity to a cliff of about 150 feet, down which
the bodies of victims may have been precipitated,
as was the custom at the inhuman sacrifices of the
Aztecs *. A road or causeway, to be noticed in
another place, terminates at the foot of this preci-

* The writings of Clavigero, Solis, Bernal Diaz, and
others, describe this mode of disposing of the bodies of those
whose hearts had been torn out and offered to the Idol.

pice, exactly beneath the cave and overhanging rock; and conjecture can form no other idea of its intended utility, unless as being in some manner connected with the purposes of the dungeon.

Hence we ascended to a variety of buildings, all constructed with the same regard to strength, and inclosing spaces on far too large a scale for the abode of common people. On the extreme ridge of the mountain were several tolerably perfect tanks, one of which approached the brow of a precipice, and was admirably strengthened in that direction, as may be observed by the diagram.

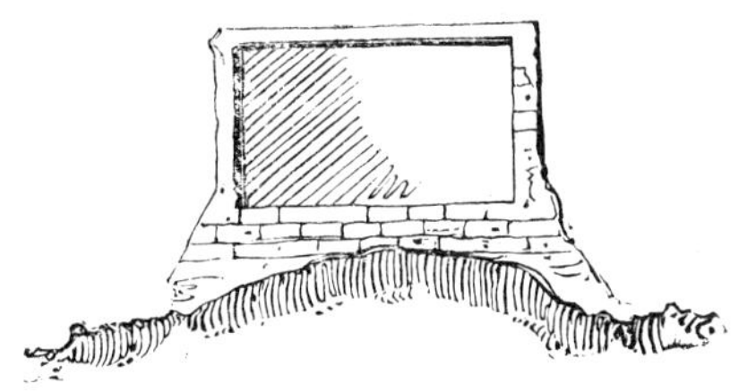

In a subsequent visit to this extraordinary place I saw some other buildings, which had at first escaped my notice. These were situated on the summit of a rock terminating the ridge at about half a mile to the **N.N.W.** of the citadel. Their disposition may be better understood by the accompanying rough plan.

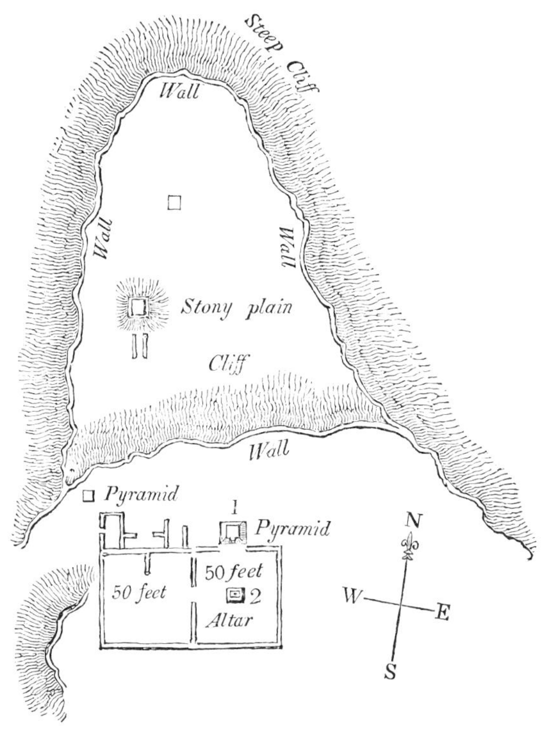

No. 1. is a building originally eighteen feet square, but having the addition of sloping walls to give it a pyramidal form. It is flat-topped, and on the centre of its southern face there have been steps by which to ascend to the summit. No. 2. is a square altar, of which the following is a sketch,

its height and base being each about sixteen feet. These buildings are surrounded at no great distance by a strong wall, and at a quarter of a mile to the northward advantage is taken of a precipice to construct another wall of twelve feet in width upon its brink. On a small flat space between this and the pyramid are the remains of an open square edifice, to the southward of which are two long mounds of stone, each extending about thirty feet; and to the N.E. is another ruin, having large steps up its side. I should conceive the highest wall of the citadel to be three hundred feet above the plain, and the bare rock surmounts it by about thirty feet more.

The whole place in fact, from its isolated situation, the disposition of its defensive walls, and the favourable figure of the rock, must have been impregnable to Indians, and even European troops would have found great difficulty in ascending to those works, which I have ventured to name the Citadel. There is no doubt that the greater mass of the nation which once dwelt here, must have been established upon the plain beneath, since from the summit of the rock we could distinctly trace three straight and very extensive causeways, diverging from that over which we first passed. The most remarkable of these runs S.W. for two miles, is forty-six feet in width, and crossing the grand causeway is continued to the foot of the cliff, immediately beneath the cave which I have described. Its more distant extreme is terminated by a high and long artificial mound, immediately beyond the river towards the Hacienda of La Quemada. We could trace the second S.S.W. to a small Rancho named Coyote, about four miles distant. And the third ran S.W. by S. still further, ceasing, as the country-people inform-ed us, at some mountain six miles distant. All these roads had been slightly raised, were paved with rough stones still visible in many places above the grass, and were perfectly straight.

From the flatness of the fine plain over which they extended, I cannot conceive them to have been constructed as paths, since people who walked barefoot and used no animals of burthen, must naturally have preferred the smooth earthy foot-ways, which presented themselves on every side, to these roughly paved ones. If this be allowed, it is not difficult to suppose that they were the centre of streets of huts, which, being in those times constructed of the same kind of frail mate-rials as those of the present day, must long since have disappeared. Many places on the plain are thickly strewed with stones, which may once have formed building materials for the town; and there are extensive modern walls round the cattle-farms, which not improbably were constructed from the nearest streets. At all events, whatever end these causeways may have answered, the citadel itself still remains, and from its size and strength confirms the accounts given by Cortez, Bernal Diaz, and others of the conquerors, of the magnitude and extent of the Mexican edifices; but which have been doubted by Robertson, De Pauw, and others. We observed also, in some sheltered places, the remains of good plaster, confirming the accounts above alluded to; and there can be little doubt that the present rough, yet magnificent buildings, were once encased in cement and whitened, as ancient Mexico, the towns of Yucatan, Tabasco, and many other places are described to have been *.

The Cerro de los Edificios and the mountains of the surrounding range are all of a gray porphyry, easily fractured into slabs, and thus with

* See the Voyage of Juan de Grijalva, in 1518 : also Bernal Diaz, Cortez, Clavigero, and others.

VOL. I.

comparatively little labour has furnished building-materials for the edifices which crown its summit. We saw no remnants of obsidian amongst the ruins or on the plain,—which is remarkable, as being the general substance of which the knives and arrow heads of the Mexicans were formed *; but a few pieces of a very compact porphyry were lying about, and some appeared to have been chipped to a rude form resembling arrow heads.

Not a trace of the ancient name of this interesting place, or of that of the nation which inhabited it, is now to be found amongst the people in the neighbourhood, who merely distinguish the isolated rock and buildings by one common name, Los Edificios. I had inquired of the best instructed people about these ruins; but all my

---

* It is not improbable, however, that this material was unknown to the nation who dwelt here, if, according to the Abbé Clavigero, this city was one of the earliest settlements of the Aztecs, before they established themselves in the valley of Mexico, near which (at Real del Monte principally) the obsidian is found in great abundance, although I believe that no traces of it are seen in the more northern provinces.

researches were unavailing, until I fortunately met with a note in the Abbé Clavigero's History of Mexico, which throws some light on the subject. "The situation of Chicomoztoc, where the Mexicans sojourned nine years, is not known; but it appears to be that place, twenty miles distant from Zacatecas, towards the south, where there are still some remains of an immense edifice, which according to the tradition of the Zacatecanos, the ancient inhabitants of that country, was the work of the Aztecs on their migration; and it certainly cannot be ascribed to any other people, the Zacatecanos themselves being so barbarous as neither to live in houses nor to know how to build them *."

---

* Clavigero, vol. i. book ii. p. 153.—Torquemuda says that the capital city of the Chechemecas was called Amaquemacan. He says this place was 600 miles distant from where the city of Guadalaxara now stands. Clavigero, who quotes this passage and comments upon it in a note, says, that "in more than one thousand two hundred miles of inhabited country beyond that city, there is not the least trace or memory of Amaquemacan."—May not the city I have described be the capital in question?

After having employed eight hours in our ramble, we rode homewards by another route, where
wonders, which in the estimation of our guides
threw the city into shade, awaited us. These were
an immense block of porphyry, called "Pie-
dra del Monarca," on which tradition reports that
Montezuma (how he came so far from home does
not appear) once reposed himself after some ar-
duous toil. There is a natural or artificial inden-
tation on the time-worn stone, somewhat resembling the print of a naked foot, which mark we
were gravely informed had been caused by actual
pressure; and although our dim sight could not
trace the outline of a monstrous hand and fingers,
corresponding but ill with the probable size of the
monarch's foot, the people saw it very clearly, and
took much pains to point it out to us.

In the evening we reached Mal Paso, and procured a room in its very miserable meson. This
place, quite a little town of itself, is the property
of a rich old gentleman of seventy, whom we saw
sitting in his balcony, with his head bound up in

a dirty handkerchief, and attended by his wife, a young woman of twenty-five. A very pretty little church stands near the Casa Grande, and has six bells mounted on a coarse series of low stages in front of it. Under a portico or gallery we saw hanging the stuffed skins of five pumas or Mexican lions, of a light dun colour; four lion cubs; twelve gray wolves and two black ones, very much resembling dogs, and which probably were a cross between a wolf and a dog *. The lions had been taken in pitfalls, and it is customary to course the wolves and laso them when heavy and inactive after a plenteous meal.

Being anxious to obtain a sketch of this place, I sat down quietly with my book for that purpose, but was interrupted by loud cries of " Pleyto ! Pleyto !" (a quarrel) which burst from the little street of herdsmen's huts. There I found three men with long knives, striking and stabbing each other most furiously, but they were soon sepa-

* In the house of Mr. Ward, chargé d'affaires in Mexico, I saw a dog and female wolf which had bred together once.

rated and led off to the Casa Grande, each wounded in different parts and covered with blood.

August 2.—Business obliged me to ride to the city on this day, to demand justice against a defaulter to the Company; and when the offender had confessed himself a rogue and bound himself to pay over the large sum of which he had robbed us, the judge, thief, plaintiff, and master of the house where the affair was canvassed, with some friends of the parties, sat down very quietly and sociably to dinner together!

August 4.—Having purchased 150 horses for the service of the mines, at twelve dollars (equal to 2*l*. 8*s*.) each, the process of casting, and branding them with the Fierro or distinguishing mark, took place on the following day, when I witnessed to perfection the whole process of the laso.

The animals having been driven into an inclosed space, it was quite extraordinary to see a small light man on foot engage in an operation which a novice would imagine must hurl him in an instant to the ground. On the contrary, the animal he selects, rushes past him at full gallop; but the noose is no sooner thrown over its fore legs, than he fixes himself in a firm position, and so resists the sudden check of the laso, that the impetuous animal is brought with such tremendous force to the ground as frequently to turn over once or twice after his fall.

This active day brought me into acquaintance with our chief Ranchero, whom I commissioned to catch some coyotes, as the chase of these animals is one of the favourite amusements of the hardy Vaqueros, who display their skill in the laso and in horsemanship in taking them. When a chase is meditated on some holiday, an old horse is killed on the plains, and a herd of coyotes soon flock to feast upon him. They are watched attentively; and when any leave the carcase, which is only when they are replete, the active horsemen enter on the chase, in which they are very rarely baffled by the wily turnings of their victims. In form, colour, and habits, the coyote closely resembles the jackal of the Sahara of Africa.

August 8.—A serious accident occurred to a poor boy, who fell from a gallery, dislocating his wrists, and otherwise bruising himself very sadly. I found him surrounded by a crowd of people, who had nearly smothered him in the folds of serapes to the total exclusion of air, taking it for granted that his recovery was impossible. I bled him, and I set his arms; and in a few days my patient was able to get about again, to the great admiration of the villagers, who were divided as to whether the cure was to be attributed to my interference, or to that of San Juan Bautista, on whom the lad had called most vociferously while I was pulling his arms.— The saint, I believe, gained the day; but my fame was also established, and my door and path consequently were beset by patients; particularly rheumatic old women, on whom I effected wonders, by the unheard-of prescription of soap and water. I visited no patients who were not properly washed;—sick men now shaved off their beards, which operation they had hitherto believed would increase their disorder; and many a poor child's head was deprived of that cake of black mud, which according to the belief of the Mexican mothers acts as a preventive to the baby "taking cold in its brains." No assurances of mine would convince the villagers that I was not a medical man, and all sorts of cases therefore came under my care. I thus obtained an influence at this retired spot, by which, in time, I might have introduced improvements in the houses and occupations of the natives, who some of them showed their gratitude for my attention to their requests by a morning present of a flower or a nosegay from their little gardens.

I really left the little village of La Sauceda with some feeling of regret, as there were many kind-hearted people there, whose character I was now beginning to understand, and which was so much more amiable than that of their Zacatecan neighbours. Two or three very worthy priests were in the habit of attending the chapel on particular days, and I conceive that to their good opinions of me, much of the civility which I received may be attributed.   I walked unarmed about the village at any hour, and never once met with an insult; and if I entered a hut, its inmates always appeared flattered by my visit.   A crowd of people assembled to give me a very sincere " Adios" when I mounted my horse to leave them; and I look back to my two months residence among them with much satisfaction.

CHAPTER IV.

Account of the Mines and Miners of Zacatecas, and Operations of treating the Ores—City of Zacatecas—Population
of the District.

THE interest which has been excited in England
respecting the mines of Mexico, induces me to
give a short account of those at which I have been
resident at this place.   It is to be observed that
the operations here have been carried on upon the
Mexican system, and without the introduction of
English improvements which have been ably commenced in other places.

The chief mineral riches of Zacatecas lie in a
cluster of high arid mountains extending about
six miles to the northward of the city, and rising
abruptly from the surrounding extensive plains.
The Veta Grande is situated nearly in the centre,
and from its superior size and great produce is
now deservedly pre-eminent. Many small mines

are dispersed around this rich vein in all directions, and their workings have been, as with the Veta Grande itself, attended with more or less success, owing to a peculiarity attendant on the veins in this district. This is their tendency to lie in rich bunches, or Bonanzas, in the ordinary course of the lode; and mine holders who have been about to abandon their undertakings in despair, have not unfrequently struck on one of these unexpected nests of wealth, which have in a few months yielded to them princely fortunes. The family of Fagoaga, distinguished latterly by the Marquisate del Apartado, are amongst the most favoured of those who have been thus fortunate, and their great riches proceeded in a principal degree from the Veta Grande of Zacatecas. This has in several instances yielded enormously, and the returns from that portion distinguished under the names of Milanesa, Urista, Macias and others in their turns, have been immense.

The Veta Grande has been worked for a vast number of years, yielding always, when not in

Bonanza, a poor but abundant ore, lying in a hard matrix of a quartzose nature, bounded by " walls" of very compact porphyry.

In width it varies from nine to sixteen fathoms, and although its extreme extent is somewhat beyond the possessions of the company, the portion in their hands runs nearly east and west to the length of 2500 varas.

The mines have been extensively worked, as from their produce and antiquity may very naturally be expected. Much however remains untouched; and when the vastness of the place is considered, I see no just reason why other Bonanzas may not, if boldly sought, be again discovered. The mines are deep, and, in consequence, somewhat wet, although they by no means bear the condemning title of " wet mines" as applied in Cornwall. The water is easy, and does not flow in any quantity or from springs, the deposit in the deep mines being rather from filtration than otherwise; increasing gradually after the rains, and in the dry season being easily kept. A great portion of the present produce therefore, necessarily goes towards the expenses of drainage, by one great shaft of *Desague,* designated as the Tiro general. To the *planes* or bottom of this shaft is 396 varas; but the water is only kept to beneath the 300 vara level, leaving the old workings for the greater part drained, but preventing the sinking of shafts on discovery, unless increased drainage is applied. It is to be seriously lamented, that the great scarcity of wood in the immediate neighbourhood of Zacatecas precludes the use of a steam power, by which the *desague* or drainage might be easily effected, and a vast expense avoided. In the present state of things, it is requisite that four malacates or horse whims should, night and day, be employed constantly, to keep the water to its present level. There are, in fact, six of these machines, the two extra ones being merely employed on cases of emergency; and all stand under three united sheds of a great size, covered with wood shingles, and called *galéras.* For the service of each malacate, fifty horses are requisite, with relays of drivers, men to attend the tackle, and receive and empty the water-skins or botas as they are raised to the surface*. The vast expenses attendant on this may be imagined; and could these be in any way immediately avoided, the sum expended on it would add greatly to the profits of the concern. A Socabon, or adit, would overcome very much of this difficulty; but it would take much time to complete this work, and objections might be made to so large an outlay as would be requisite for its accomplishment, although it is evident that its expense would be far less than two years' *desague* by the present process.

* The total number of horses thus employed, — the mules at the Veta, and those for the Arastres or Tahonas at the Hacienda, the asses which carry ore, and the horses employed to tread the Tortas, — exceeds 1200. It is in consequence of the fluctuation in the price of fodder for these animals, that no correct annual estimate can be made of their cost. A dry, or even too wet a season destroys the crops of maize; and grain which in one year may be procured for seven reales the fanega, can scarcely be obtained in the following, under twenty, thirty, or even more—so that in so large an establishment there may in some years be an increase of expense in this article alone, to the amount of 30,000 dollars.

The workings of the last possessors of the Veta Grande, previous to its delivery to the Bolaños company, were left in many instances in so choked a state, that ventilation was much impeded; and time and labour have been absolutely requisite to restore this important object, by clearing the levels, and in other ways putting things in order. A mine which has been carelessly worked, as far as air, order, and facility of access go, presents obstacles to those whose object it is to remedy these evils, which cannot be imagined by persons unacquainted with mining operations; and in the Veta Grande much was required in this way, and much has been accomplished.

Many of the dressing-floors situated near the mouths of the principal shafts were easily accessible to plunderers, from the inefficient state in which their walls were left—some had no walls at all. These are now properly inclosed; the dressing-floors are levelled and in better order, and much improvement is perceptible in the surface work of the mines.

The process of breaking, selecting and dividing the ores, as delivered immediately from the respective workings, is performed by men under the superintendence of native ore-dressers, who keep an account of the quantities received, their weight after selection, and of their delivery to the " Conductor," whose duty it is to carry the ores to the Hacienda of La Sauceda.   For this purpose there are two hundred fine asses, which are each laden with two leather bags, containing together one carga, of three quintals, equal to 300 pounds. The bags are weighed at the Patio, or dressing-floor, registered, and an account is sent to the Hacienda, whence a receipt, on their being again weighed, is returned. The process of amalgamation, as practised with such success and superior expedition (compared with other mines) at the Sauceda, is most fully detailed, in every one of its stages, in a paper which I wrote after two months study of the subject, and which is in the Appendix.

The Hacienda, as before stated in my journal,

is about five miles to the N.E. of the Veta Grande, so near the foot of the mountains that laden waggons approach it with the greatest facility from the adjoining plains, and a command of water for the purposes of amalgamation can constantly be obtained by means of a *noria* or water-wheel.

The bars of silver are sent from the Hacienda to the Veta on the Saturday, thence on Monday they are forwarded to the Mint at Zacatecas, (which is unquestionably the most effective one in the Republic of Mexico;) and on Friday of the same week the coined dollars are returned to the Veta, so that they can be employed in the payment of the labourers and other expenses on the Saturday, only one week from the time of casting the silver into bars. The expenses incurred in the mint will also appear in the Appendix, showing the total cost of changing the metal into currency.

Here it may not be amiss to give some description of the Mexican, or perhaps more properly, the Zacatecan Miner. These people are not, as is generally supposed, Indians. I seldom saw a

pure Indian in a mine in this part of the country,
but they are generally tinged in some degree with
Indian blood, although I conceive that their affi-
nity to the Whites is somewhat the nearest. They
are in fact mostly Creoles, of a deep tawny hue,
with jet black glossy hair, dark eyes, and a pene-
trating expression of countenance seldom met with
in the Indian.   In stature they are of the middle
size, usually of a spare muscular habit, active and
enduring in labour, and when sober, or not under
the influence of the intolerant feelings which are
in this district so visible towards foreigners, are a
lively, somewhat tractable race.   When inflamed
by liquor, or their passions are excited, they are
violent, bloody, and revengeful. In time, however,
I conceive that their character will change very
materially for the better; since I believe sincerely
that, if unurged by some few bad and more intelligent spirits than themselves, they are capable of
attachment, and have many good qualities. Their
wages are high, and in consequence they are im-
provident; spirits, the universal passion for gam-

bling, and the extravagant dresses of the females of their family, unite to swallow up their well-earned wages, and the poverty incidental to their want of care and foresight is never attributed to the true cause. The mendicant friars also deprive them of a considerable portion of their pay; as on the settling days, large subscriptions are made at the pay table for the good of various religious establishments.

The climate at the mines is extremely healthy, the temperature mild in summer and somewhat cold in the winter, varying in each a few degrees from that of the Hacienda, where during the two months of July and August the thermometer ranged from 64° in the night, to 74° at mid day.

I am sorry that it is not in my power to say much in favour of the city of Zacatecas, which I believe was once the capital of a powerful nation, (the Zapotecas,) who were subjugated with great difficulty by the troops dispatched by Cortez for that purpose after the conquest of Mexico*. I ac-

* Vide Bernal Diaz.

knowledge a dislike to both the natives and the town, which I only entered five or six times on business; and I had no idle time on my hands, had I been disposed to make my visits more frequently. Thrice I so far succeeded in attracting public attention as to be hooted at as a Jew, and once had the honour of being pelted with stones. The frequent use of the knife is also a sufficient discouragement to a stranger's visiting the city. Murder is too slight a crime to merit punishment,—and during the month of May, twenty-one assassinations took place without a single person being brought to justice.

The town itself is good, but from the inequalities of the ground the streets are short, uneven and crooked. In some of these are foot pavements, and the place generally speaking is clean. The churches are large and very well built, and the Parroquia (the parish church) is certainly a noble edifice. Its front is superbly ornamented, and entirely covered with rich carving in stone; the architecture of the belfry is beautiful. Its font is one of the wonders of Zacatecas, being entirely of silver, and weighing 3793 ounces. The execution however is greatly inferior to the material. " This baptismal font was presented on the 20th of November 1800, by Doña Maria Anna de la Campalos, countess of San Matteo Valparaiso, in remembrance of her having received the waters of holy baptism in this church, under the condition that if any other should present a better font, this shall be removed to the church of Sombrerete. The weight of this font is 474 marcos and one ounce." The above is engraved round the margin of this ornamental " Pila," which stands in a small room tawdrily painted in fresco, and bearing on its walls a variety of most extraordinary verses in a doggrel style, which I am not sufficiently skilful to translate.

The market of Zacatecas is tolerably supplied with fruit and vegetables, chiefly the produce of the Barrancas near Guadalaxara; but the natives consume so little of the latter compared to Europeans, that the supply exhibited in the morning would appear insufficient for one of our small villages. The principal, and to us the most interesting building, is the Mint, which has recently been put in very excellent order at considerable expense, and is unquestionably the best of the five in the Republic. Three excellent dies can be kept constantly at work, and each averages, if strictly attended to, forty-seven dollars a minute; which, allowing eight working hours to the day, gives 22,560 dollars. The milling process is ingeniously arranged, and the casting is the only defective part of the establishment, which is able on receiving the bars of silver from the mines on the Monday, to deliver their amount in dollars by the Friday following,—a very great advantage to the proprietors; whereas in the Mint of the city of Mexico, a much longer period must elapse before the returns are made. La Casa del Ayuntamiento (or del Estado,) is really a magnificent building, perfectly clean and well ordered. Here all the public offices are established, and the sovereign congress of the state (Soberano Congreso) assemble.

There is no lack of friars of various denominations in Zacatecas, of which the Blue Franciscans are considered the most dissolute. Their respective convents are, as usual, the finest buildings in the place, and their influence, although somewhat on the decline, is still very extensive.

The governor resides in the city, as does also the general of the state, to whom I paid a visit. There are no troops in the town beyond the " Milicia," a circumstance on which the inhabitants pride themselves not a little; but which is to be deplored, as one reason for the state of disorder and anarchy which so frequently reigns; for in other towns where there are garrisons, better order is preserved.

No manufactures of any importance are carried on here, with the exception of a government establishment (Fabrica) for making cigars, of which the consumption in every part of the country is almost incredible.

Foreign merchandize is to be procured from several very good shops, but is expensive on account of the distance of its transport; and the only commerce of Zacatecas is with the neighbouring towns, of which, as the capital, it is the focus.

When viewed at a mile distance, either from the north or westward, Zacatecas has a most beautiful and imposing appearance, lying at the foot of an abrupt and picturesque porphyritic mountain, named the " Buffa," whose rugged summit is crowned by a neat church and a small fortress which was erected during the revolution. There is a narrow Alameda attached to the city, border-ed by a row of young trees, the only ones which the natives have been at the trouble of planting, although very little care and expense would quickly clothe many parts of the mountain with timber, and remove that air of desolation and barren-ness which surrounds "the mother of the mines." A quantity of small huts are scattered about near the city, of which the population, exclusive of these mining villages, is said to be 25,000; but this I consider as about double the truth. The country is healthy and the climate pure and agreeable, averaging during my residence, a general temperature between 64° and 74°, but being from its elevation very cold in the winter season. The sky is here usually very clear and brilliant, with the exception of the season of the periodical rains, and the natives live to an advanced age, if not of that class attached to the mines, among whom the free indulgence of spirits must tend considerably to shorten the period of existence.

The people of the mining districts have the character of being more lawless and unruly than those whose occupations are different; and whatever may be the truth of this imputation as regards other mining states, the Zacatecanos are somewhat worse than their neighbours. I do not however conceive that the mining interests of foreigners can now be materially or even slightly affected by the waywardness of the operatives. Mexico is a country newly awakened from a long dream of ignorance and oppression; and as much improvement is already observable to the residents in the country, more may naturally be anticipated, although its progress must, I conceive, be slower

in the state of Zacatecas than in the more central
provinces, since the natives possess more bigotry
and intolerance than their neighbours; and any
improvements introduced by men of a different faith
from themselves will for a period be received with
distrust, and were at first exposed to insult. It will
scarcely be believed that there should exist a people
in a nominally civilized country, who yet believe
in Lord Monboddo's ingenious theory of tails,—yet
so it is; that the English, or indeed all foreigners,
being considered as Jews, are supposed to be ornamented by these appendages; and many people
can be found who firmly believe that our stirrups
being placed more forward on our saddles than is
the custom of the country, is to allow of our stoop-
ing a little so as to prevent the friction of the sad-
dle from inconveniencing the rider's tail.

It is to this bigotry that the circumstances of in-
sults with which some of our people were received
on their first arrival, are to be attributed. The
prejudice of the people, influenced by the ignorant
priesthood, induced them to look with jealousy

upon all foreigners as heretics.   This prejudice is greater in these northern states than in the other parts of the Republic, and may be attributed in a great measure to the little intercourse they have had with Europeans, and will wear off gradually with the general improvement which this country must experience.   In other parts of the Republic our countrymen have been well received.

CHAPTER V.

From Zacatecas to Bolaños—Villa Nueva—Native Tea—
Warm Springs at Encarnacion—Manufacture of Cigarros
—Colotlan—La Aguila—Town of Bolaños—Rope Dan-
cers—Guichola Indians—Ball.

August 18.—HAVING made the necessary arrange-
ments in regard to my duties at this place, I now
commenced my tour to Bolaños and the other establishments of the Real del Monte and Bolaños companies, previous to my return to England ; leaving
my friend Mr. Tindal, who for this purpose with
great kindness agreed to postpone for a few months
his return to England, to occupy my place, in charge
of the Hacienda, while Dr. Coulter remained in
management of the mines of the Veta Grande.   I
took leave of the Hacienda in the evening, and
sleeping at the Veta, left it on the afternoon of the
19th of August.   My party consisted, besides myself, of two native servants, two arrieros, some

laden mules, and a few spare saddle-horses, with which I was about to traverse between three and four hundred leagues, before my embarkation for Europe. We reached Mal Paso soon after dark, and thoroughly soaked by the rain, an inconvenience for which we could find no alleviation in the comfortless Meson.

August 20.—I set out for Villa Nueva, and on my way devoted half an hour to the examination of a portion of the Edificios, which had before escaped my notice. In riding slowly up the rugged side of the mountain, my horse suddenly made a full stop; and I saw immediately before him a large rattle-snake with open mouth which appeared disposed to dispute our passage. Wishing to obtain this reptile as a specimen, I was rather careful in killing him with stones, and therefore had a good opportunity of hearing his rattles to great advantage; ascertaining also that the story of three warnings could not in this instance be relied upon, since the rapid vibration of the tail continued unabated until the creature was killed. Its colour was much lighter than that of those exhibited in England; and although it had eight rattles, the length was only three feet eight inches. The fangs, however, were very large, and I dropped a dollar through its mouth with the greatest ease. It is generally believed by the natives that the bite of the rattle-snake is rarely fatal, and that a solution of nitric acid applied to the wound as a lotion is an effectual cure.

Passing through the Quemada, at the distance of eight leagues from Mal Paso, and fifteen from Zacatecas, we arrived at the pretty little town of Villa Nueva, where with all our live stock we became the guests of Don José Maria Marques, an old friend of our *Negociacion.*

Our ride from La Quemada had been over a smooth grassy plain, whence by a gentle descent we entered a narrow valley abounding in flourishing little gardens and plantations of maize, and passed into the town through a small grove of fresh-looking poplars, beneath which some lively and well-dressed groups of country-people were making merry.

In the evening a procession of some celebrated

Virgin moved through the town, preceded by fid.. dles, fifes and guitars, and followed by a prodigious number of women. All the people who did not join in the train stood at their house doors, and we knelt until the idol had passed. The whole ceremony was conducted with great solemnity until a loud screaming announced the arrival of an enraged cow, which dispersed the worshippers in all directions, and drew the attention of the young men from more serious subjects to their favourite amusement called Colear, which consists in trying to throw cattle by a peculiar manner of catching them by the tail,—an operation which soon drove the intruder bellowing down the street.

My hostess presented me for supper with a cup of native tea, which some stranger, whose name or nation I could not understand, had gathered on a Rancho, the property of Don José, and where he said the plant was to be found in great abundance. In the form of the leaves and flavour of the infusion it struck me that this was a species of the China tea-plant, and I advised Don José to make diligent search for so valuable and important a production.   I very much regretted my inability to visit the Rancho whence the tea had been gathered, as also to extend my trip about three leagues to the S. W., where is situated the Hacienda of Encarnacion, the property of the Marques del Xaral,—one of the richest proprietors in the Republic of Mexico, and celebrated for the superior breed of horses which he possesses on his estate.

Encarnacion is interesting, as having some warm springs of a most agreeable temperature for the bathing of invalids.   The waters are said to contain nitre and lime in solution.

At about five leagues to the eastward are other baths, also warm, and strongly impregnated with sulphur.   These belong to a Hacienda called " Tepetistaque," through which also runs a stream abounding with a fish called Bagre, resembling the catfish of brackish rivers.

August 21.—The town of Villa Nueva is neatly built, possesses some good shops, and has a population, according to Don José, of 6000 souls.   It is one of the dépôts of tobacco, which under the new as well as the old regime is a strict government monopoly. While the mules were being saddled, Don José very obligingly accompanied me to the "Fabrica,"—a large well-arranged house, in which 400 men and 350 women are constantly employed in the manufacture of "Cigarros." This is the name given to those formed of cut tobacco enveloped in paper, while the term "Puros" is applied to the rolled tobacco leaf which in Europe is commonly called a cigar. Distinct portions of the house with separate entrances are appropriated to the sexes, who are distributed in long rooms having several rows of benches. Each labourer has a small basket with a certain weight of rasped tobacco, and sufficient papers ready cut to contain it when made into cigars; and when this proportion is disposed of, it is rigorously weighed and registered. From three to four reales is the average price of a day's labour, which commences at 5 A. M. and ends at the same hour in the afternoon. The expedition with which some of the most active people rolled the cigars was quite extraordinary, and there are many who complete 4000 in a day. The product of the last four days and a half had been 121,309 " Cajas" or paper parcels, each containing thirty-two cigars, making a total of 3,881,888 ! the expenses of working which was 1115 dollars. The cajas are sent to the market packed in chests, each containing 4300. The distribution of labour at this establishment is very well arranged: from the makers the cigars are carried to the counting-room, where they are expeditiously made into cajas, and pasted in a paper bearing the stamped seal of government. The work-people are strictly examined, that they neither introduce liquor or weapons, and both sexes are searched nearly to the skin before retiring for the night, for which purpose female searchers, " Registradoras," are stationed at one door, and men at the other.

At eight I set out on my journey, but not until I had received a large lamb, roasted whole, from my hospitable friend, who assured me that I should otherwise fare very ill on the road. The morning was delightfully fine, and we passed for about five miles over a plain thickly covered with young maize, amongst which the countrymen were turning up the furrows with small light ploughs drawn by two oxen, with their heads tied up very high to prevent their eating the fresh green blades. From hence we entered amongst the mountains, where for the remainder of the day our road was very bad. At seven leagues, having toiled up the Cerro de Membrillo, we descended that of Huacasco, by a road not easily forgotten, to a deep wild dell, in which we stopped for a time at the small Rancho de Huacasco. Here we ate Tortillas and drank whey in the house of a poor woman who had never in her life seen a foreigner, nor moved one league from the spot on which she was born. I gave her a fine brass ring as a souvenir of this visit, and having told her I was "Yngles," received with her thanks the assurance that she had never heard of such people.

In a ride of three leagues over the mountains, the road descended to a small fertile vale, rich in maguey and maize, near the pretty little shady Rancho of Tenasco. Hence again entering on the mountains, we did not reach the village of Colotlan until long after night-fall, having occupied eleven hours and a half in riding fourteen leagues. The day having been very sultry, my animals were much fatigued; yet we could find neither food nor a resting-place for them, it having happened that a rope-dancer had recently arrived from Guadalaxara, and all the Colotlan world who could muster three-pence had flocked to see his performance. Men and cattle, therefore, went supperless and comfortless to bed. The mountains amongst which we had this day travelled, were of red porphyry, coated in many places with chalcedony; and a peculiarity of form was observable throughout, or with very few exceptions. They were for the most part crowned by an abrupt naked ledge of rocks resembling a wall; above which was a space of perfectly level land, whereon trees were very rarely seen growing. The lower ground on the contrary was thickly clothed with stunted oaks, (amongst which I observed one species resembling the *Quercus diversifolia*, mimosas, and tuna. The whole country was abundantly covered with fine grass, offering a most striking contrast to the brown arid deserts around Zacatecas. Amongst so much verdure I could not but observe the devastation which the numerous communities of ants had caused, by clearing the ground of every vegetable substance to the diameter of six to ten yards around their thickly-peopled mound, which was usually elevated about a foot above their circle of desolation.

August 22.—Being unable to procure food for my hungry cattle until late this morning, I was obliged to give up all idea of travelling further till the cargo mules had been refreshed. I therefore occupied the early part of the day in rambling by the banks, and bathing in the small turbid river which flows through the valley of Colotlan, and affords with its attendant gardens and fruit-trees some delightful scenery. Maize is here extensively cultivated, and the maguey grown in considerable quantities. Being lodged in the room of the

Ayuntamiento, or Town-hall, adjoining to the common gaol, my windows were constantly filled with gazers from the crowd, who in the Mexican villages are always to be seen assembled round the prison bars, where all the gossips concentrate, and all important village questions are discussed.

In the evening I received a visit from a very great man, the Gefe Politico, whose duties I could never clearly understand; but he brought with him a Licenciado, a Secretary, and a numerous unshaven train of attendants, with whom I had a long conversation upon the subject of the penal code of Xalisco*, which, as well as the trial by jury, my visitors informed me was copied from the laws of England. The criminal code of Zacatecas was spoken of with high contempt, as having no definite punishment for murder; while that of Xalisco was vaunted as a pattern to the Republic. I afterwards found that this was not exactly the case; that not a single law is ever put in force, and that

* The State of Guadalaxara.

the impunity of crime at the one place fully equals that of the other.

While we were debating on these important subjects, the sensitiveness of the chief authorities as to criminal matters was put to a sufficient test by a loud outcry from my neighbours in " durance vile," who were fighting with knives, and had wounded one of their number. This was an admirable foundation for a display of justice: but alas! cutting and maiming are not mentioned in the penal code; and the gaoler very jocosely made his report of the " Cuchillada," at which the au·· thorities all laughed, and then continued their conversation. The most intelligent of my visitors was a pure Indian, proprietor of a Rancho, and of the Tlascalteco nation, who with some Chichimecos de Soyatitan and Tochopa Tepehuánes are established in the town and its neighbourhood.

Spanish is the only language now generally spoken by these people, although many still retain in their domestic circles the dialect of their ancestors. There are in Colotlan and its vicinity (all of whom come in to the Sunday's mass) 7000 souls, governed by the Gefe Politico, an Alcalde (who is an Indian), an advocate, and two Alcaides to the prison; from which strong hold I was several times accosted by a noisy fellow, with "I say, John English—ha ha! my boy!" which, unfortunately for further conversation, was my friend's whole stock of English.

August 23.—Our road, although mountainous, was good throughout the day, and maize fields were frequent. In five leagues we reached a retired shady village of a few huts named Cartagena, standing at the head of a small valley, through which runs a picturesque little river of the same name. Its course at this place was to the northward, where making a short turn it is said ultimately to connect itself with the river of Bolaños. We forded the stream, which at this time was little more than two feet in depth; but it is said to increase very much after heavy rains, when its course is impetuous, and serious accidents frequently occur in passing it. At these times passengers are

slid from bank to bank upon a rope, and animals
are conducted across by dexterous swimmers, who
are paid very highly for this service. Here I
drank some excellent Pulque under the shade of
a fine Mesquiti tree, from which station I could
keep a watchful eye upon my cargoes, as well as
admire the surrounding scenery,—the natives of
Cartagena being celebrated for their ingenuity in
depriving a mule of its load. We now immediately
ascended to the higher land, where by the road
side we passed two well-dressed but suspicious-
looking men, seated amongst the bushes; and evidently, from their manner, on the look-out for ourselves or some other travellers. They followed us
for some little distance, having their Serapes so
disposed in folds about their persons that we could
not see if they carried arms: but either disliking
our number, or being disappointed in others who
were to have joined them, they soon disappeared
in the mountains. Passing several small Ranchos,
we ultimately ascended a mountain clothed with
three species of dwarf oaks, one resembling the

Ilex in its leaf; a second being gnarled and tortuous, and with leaves like our English species; the third was straight and slender in limbs and branches, averaging about twenty feet in height, but bearing leaves which were almost all from twelve to sixteen inches in length: specimens of these I have brought to England with me, one measuring nineteen, another twenty-one inches in length. I lamented that no acorns were at this season on the trees; but many of those of the last year lay on the ground, half-decayed and worm-eaten; they were not larger than our ordinary kinds in England. This species is, I conceive, a variety of the *Quercus macrophylla*, or Large-leaved Mexican oak, described by Louis Nee* as having the leaves a foot long.

The mountains are of a compact white freestone, much discoloured at the surface, and strewed in many places with pieces of porphyry and fragments of black lava. The pass is about three miles in length, after which, leaving a rude woody

* See Rees's Cyclopædia, article "Quercus."

gorge with high cliffs on our left, we descended between steep rugged rocks to a very smooth and verdant plain, in the centre of which are the little huts of Salitre*. In one of these I slept, and early on the following morning (August 24) set out for Bolaños. The plain was cultivated for a few miles with maize, now in a very forward state, amongst which I had two ineffectual shots at a distant deer, and having advanced nearer was about to take more certain aim, when a man galloped up to me at full speed, declaring that the creature was tame; and to exemplify this, he turned his horse's head, when the pretty animal followed him playfully at a rapid pace across the country. About four miles from our outset we entered the mountain passes, which are on a most magnificent and beautiful scale; and for several hours we continued winding amongst close thickets of the broad-leaved oak, and a species of mimosa, bearing a small yellow puff or flower which gave forth a most delicious fragrance. Late in the afternoon we reached the

* Fourteen leagues from Colotlan.

extreme crest of the mountains, and on rounding a little flowery bank a scene of indescribable grandeur presented itself. To the left is a dizzy cliff, La Aguila, 1500 feet in height; its summit dazzling beneath the rays of the sun, while its outline was finely marked by a dense mass of thunder-clouds in the distance behind it. A deep woody dell crossed from what then appeared to be its base, to a smaller but no less beautiful preci‑ pice which bounded the view to the right; and the dark border of this connecting wood was sharply drawn against a purple and distant range of moun- tains to the westward, fringed on their summits with towering pines, which stand at an elevation of 5000 feet above the Barranca of Bolaños.

Our road lay along the foot of La Aguila until we reached the crest of the mountain which hems in the woody dell, whence a glorious view of the valley or Barranca of Bolaños, at a depth of 2000 feet, burst suddenly upon the sight; its shining river running through the beautiful vale, and the Hacienda of Tepec just showing the tops of its buildings from amidst the brilliant foliage in which it is embowered. We occupied nearly two hours in the difficult and steep descent, under the shade of a delightful wilderness of trees, of species quite new to my eye, and twining to the left. After arriving on level ground we soon reached the town of Bolaños, where in the excellent house of Mr. Auld, who was in charge of the concern, I found a most kind welcome.

Having never had the luxury of an entire room to myself since arriving in the interior, I seemed now to have entered a palace; indeed the whole of the slightly-peopled town bears every appearance of having once been of the first order: the ruins or half-finished remains of splendid churches and fine buildings of freestone were equal to any thing I had hitherto met with. There was not a single mud hut or hovel in the place: all the dwellings were built of stone, in a superior manner; and the public edifices now untenanted, the ruins of immense Haciendas de Plata, and other establishments attached to the mines, all bespoke the immense wealth and splendour which must once have reigned in this now quiet and retired spot.

August 25.—Mr. Auld accompanied me at an early hour, in a delightful morning, to visit the Leet Head, or that part of the river which is to be turned into the canal for the purpose of driving the water-wheels of the great mine of " Barranco." This leet commences about four miles to the northward of the town of Bolaños, and the surrounding scenery is really superb: abrupt and dizzy precipices of freestone confine the valley to the eastward, and masses of picturesque hills are covered with verdant and flowery shrubberies; this being the spring-time of the Bolaños year, all nature looked so fresh and beautiful, that the sultry heat was almost forgotten amidst the bright young foliage, the gurgling of the river, and the sweetness of the air. It is during, or immediately after the periodical rains that vegetation appears to take a new impulse; and the face of nature undergoes a more striking change than is to be seen in climates where the showers occur at irregular intervals throughout the year. In the "dry season" all the surrounding country appears burnt up and parched; no bright blade of grass or lively flower refreshes the eye, and the foliage of the trees appears, although green throughout all seasons, of a dull unrefreshing hue. I here took some sketches, though with a full sense of my inability to give more than a very faint idea of the country around me. In the evening, Mr. Price took me through a shady ride at the foot of the mountains to a little farm named " Comite," situated in a woody dell, not far from a delicious basin-shaped valley, which resembles some of those seen amidst the Alps, and has not unaptly been named " The Vale of Peace." The paths through which we rode resembled some of our ornamental coppices in England, and are continued for many miles along the depths of the Barranca.

August 26.—My morning was devoted to the examination of the workings upon the leet; and by descending eight of the ten lumbreros, or air-shafts, I became enabled to judge of the very great progress which had been made in this important undertaking.

As this was the pay day for the native labourers, the evening was in consequence devoted to merriment, which was quite a contrast to those Saturdays meetings which I had seen at Zacatecas.—A party of itinerant Maromeros (or rope-dancers) held their exhibition in the large walled yard of a once splendid mansion, to about eight hundred people; which was considered as a very " full house," the receipts, at a medio (or three-pence) for each person, amounting to fifty dollars. The performance, which was exceedingly bad, was nevertheless highly applauded by the spectators, who were sitting or lying in a confused multitude on the bare ground; while some few persons of distinction had taken the precaution to provide themselves with chairs and stools. During the exhibition of the tight rope dancers, the spectators derived a continual fund of gratification from the Pallase or clown, who particularly delighted the most respectable inhabitants by the recital of a coarse story.   While a half-Indian was performing some clumsy evolutions on the rope, the band, in obe- dience to a signal, suddenly ceased; and the dancer having dropped himself into a sitting posture on the cord, pulled off his high embroidered cap, and very gravely thus bespoke us: "Caballeros y Señoras (Gentlemen and Ladies), I beg (suplico) that as I am about to throw a somerset, you will subscribe some money to be devoted to the service of celebrating the holy sacrament of the most Holy Mass." All rose:—the men took off their hats with the utmost gravity;—a general silence prevailed for a moment; and the vaulter, who evi- dently was in great dismay, attempted to throw his promised caper. Unluckily, however, he failed, tumbled on his nose, and no money was subscribed for the " solemn and most holy sacrament, " for- asmuch as the articles had not been fulfilled. To this succeeded fireworks, tumbling by two little boys, and performances on the slack rope; in which the unsuccessful vaulter astonished me by hanging with the rope at full swing and high above the ground, by one hand, by his heels, his toes, the back of his neck, and lastly by his teeth. He concluded with a performance which is said to have been exhibited by order of Montezuma for the amusement of Cortez and his officers; and which I cannot better describe than in the words of the Abbé Clavigero, substituting however *boys* for *men*. " One man laid himself upon his back on the ground, and raising up his feet, took a beam upon them, or a piece of wood, which was thick, round, and about eight feet in length. He tossed it up to a certain height, and as *it* fell, he received and tossed it up again with his feet. Taking it afterwards between his feet, he turned it rapidly round; and what is more, he did so with two men" (boys) " sitting astride upon it, one upon each extremity of the beam." The feat, however, was in the present instance accompanied by a lively tune from the band, to which the performer kept excellent time, while he danced, with his feet elevated beneath the beam, a very neat and difficult figure throughout the exhibition.

While all these gaieties were going forward, two or three men constantly occupied themselves in picking their way through the crowd, and bawling lustily "sweetmeats and cakes for sale:" and one old fellow particularly pleased me, by his energetic yet conciliating appeals to the gallantry of the gentlemen present, to purchase a kind of "Pan dulce" which was squeezed into the semblance of pigs—"What! Caballeros! does no one buy my pigs for the ladies? What! no pigs for the señoras?" an appeal which had such effect upon the Bolaños beaux, that many a fair mouth soon blew forth its cloud of smoke, relinquished its cigar, and swallowed a "puerco."

Our evening's entertainments,—all for the price of three-pence,—were concluded by two comedies in front of three sheets, which performed the part of scenery. One was tolerably good, being a mutilation of Molière's "Mariage Forcé:" the other, which was highly applauded, I will not describe. The spectators, although a parcel of Indians and half-casts, the greater part without shirts, would

have taught a lesson of quiet and good-breeding
to our London audiences, much as we pride ourselves on our superior politeness and decorum; I
never indeed saw so large a body of people more
perfectly well-behaved, silent, and good-humoured.

August 27.—Sunday being the market-day at
Bolaños, the little square in front of Mr. Auld's
house was crowded with people at a very early
hour; and about twenty of the Guichola Indians
(of the same race as those seen by Captain Basil
Hall at Tepic) were amongst the traders, selling
a coarse kind of salt which they had brought from
the shores of the Pacific. Each man carried his
short unornamented bow in his hand, and a well
stocked quiver of deer- or seal-skin at his back,
while some also had two or three loose arrows
stuck through their belt. These arrows are of
light slender bamboo, generally fitted with a long
point, of some hard wood, yet a few were headed
by a thin small piece of copper. The dress of the
Indians was principally of a coarse blue or brown
woollen of their own manufacture, formed into a

short tunic, belted at the waist and hanging a little way down before and behind. Many had no other clothing of any kind; but the breeches of the few who wore them, were of ill-dressed deer- or goat-skin, deprived of hair, and not even descending to the knee. At their lower edges are strung a quantity of slender leather thongs, which are said to contain an inventory of their goods and chattels, including wife and children. After some hours fruitless endeavours to purchase a pair of these singular articles, I at length succeeded in obtaining a very ragged and greasy pair, with which the owner parted most reluctantly, as they bore the register of his cows, and bulls, and calves. For my own part I could perceive but little difference in the appearance of these thongs, except some irregularities of length; but there seems no doubt as to the fact of the Guicholas keeping an account of their property in this peculiar manner: Captain Hall received the same account of the knots of these inventorial breeches. The men wore round the waist or over their shoulders several large woollen bags, woven into neat and very ornamental patterns, and in which they carried their food, money, or purchases at the market. All the married men wore straw hats of a very peculiar form, with wide turned-up rims and high pointed crowns, which near their tops are bound round with a narrow garter-shaped band of prettily woven woollen, of various colours and having long pendent tassels. These people cherished a profusion of bushy black hair, in many cases confined tightly round the crown of the head by a band similar to that which encircled the hat; and almost every man wore an enormous pigtail, bound up in other bands, having large heavy-looking tassels, which generally descended below the waist.

I was informed that no unmarried man or woman may wear a hat, or bind the fillet round the head; and as we saw some young people who had neither of these ornaments, it may, in all probability be the case. There were two young married females of the party, each wearing a hat similar to those of the men; and one of them had her head

ornamented with a scarlet band. Two of the men
and one woman came into the house that I might
sketch them. They scarcely understood even a
word of Spanish, but fully comprehended what I
wanted, and were very quiet and good-natured. The
girl wore an immense roll of white beads round
her neck, and from each ear a long bunch, from
which was suspended the half of a little cockle-
shell. Her shoulders and body were covered by
a rough coarse cloak of brown woollen, without
sleeves, having merely a hole through which the
head was put; and she wore also a petticoat of the
same material, barely reaching below the knee :—
she was, as were her countrymen, barefooted;
and I observed that the great-toes of all these
people were much more separated from the others
than is the case with Europeans. In complexion,
feature, hair and eyes, I could trace a very great
resemblance between these Indians and the Esquimaux, who are, however, somewhat shorter and
more corpulent. They are said to be a very peaceable inoffensive race when sober, but quite out-

rageous in their drunken fits, when their quarrels are very bloody. Their marriages are curiously conducted: since it is the custom for a man to take his intended wife on trial; and if, after an indefinite time, he likes her, they are then married by a priest or friar, who once a year goes round to perform this ceremony, and to christen perhaps the offspring of half the newly married couples. Should the lady not give satisfaction, she may be returned to her parents, even if pregnant; and women who have been thus discarded, are as frequently taken again on trial, and ultimately married, as any others.

The Indians were not the only novelties I met with on this day; for I was made acquainted with a most extraordinary man named by the natives "Don Justo," and who, for some unknown reason, has not worn clothing or slept under a roof for many years. Round his waist he was girded by a kind of kilt, composed of many hundred little strips of rags strung and matted into a thick mass. From his left shoulder, and crossing to his right side he bore, in the manner of a knight's ribbon, an infinite number of little coils and bunches of small rope and twine tied and twisted together, and round his ancles were hung quantities of little straps and pieces of leather, in such a manner as to cover his feet entirely beneath two bunches resembling mops. The rest of his person was completely naked. This singular man possessed an intelligent physiognomy, was quiet and unobtrusive in his manners, perfectly rational in conversation, and never begged, although he would receive, in charity, whatever his few wants required. The general idea respecting his continuance in this miserable state is, that in consequence of some disappointment in love he had bound himself by a vow to his present wretched life. I made a sketch of him, which in point of novelty and peculiarity may vie with the portraits of any of the worthies in the " Wonderful Magazine."

August 28.—My day was again employed in the underground examination of the mines ; and in the evening I was invited to a ball, for which a party of ladies were assembled in a line of battle at one end of the room, and keeping up a constant fire with great seriousness and precision. The rope-dancers were of the party, and performed the " Xarabe," a kind of fandango, of which the na-tives are passionately fond. It consists of variously timed shuffling with the feet, the knees being bent, the body perfectly erect, and the hands dangling by the sides, as with sailors in a hornpipe. The dance, or rather the very dull music, is interrupted at intervals by a monotonous chaunt from the dancing woman, or a volunteer from the company, whose chief merit as a singer appears to consist in the excessive shrillness of the tones which she produces through her nose,—a musical peculiarity distinguished by the name of " Gaugoso." All the singing, in fact, amongst the lower orders in Mexico is strictly nasal; and even duets and trios by ladies of the middle classes are not in parts, or variously modulated tones, but all the perform-ers sing at once, as loud as they can, in the same twanging disagreeable voice, which is considered as very fine.

There was some bad dancing, by really a very nice-looking set of young women,—without stays of course, but very neatly clothed. Great decorum was maintained throughout the evening; and one end of the room was crowded by Leperos in their blankets, who very quietly squatted down to see the ball. The clown of the Maromeros, a well-looking man now that the black paint had been washed from his face, was much pressed to join the company; but the poor fellow modestly stood at the door, and ruefully shaking the sleeve of his clean but ragged shirt, declared himself unfit for such good society; although some muleteers and one of the servants were figuring away at a great rate. Some of the matrons, much to their credit, had brought their tender charges with them; and at intervals between the dances all the duties of nursing were performed, and the babies again deposited to sleep in the various corners of the room.

August 29.—This morning I visited the celebrated mine of " Barranco," and saw such of the old workings as are still clear. At this period nothing was to be brought away as a remembrance of this place, but half-a-dozen bats, of which immense flights now occupied the shafts and levels whence millions of money had once been extracted. In the evening I forded the river on horseback, to sketch the fine extensive ruins of the surface works of this great concern, which from their superior masonry and regularity of structure must have once resembled an immense fortress.

August 30.—I devoted my forenoon to visiting a very fine unwrought vein of rich copper ore, situated at the northern extreme of the Cañon of Bolaños, at an elevation of about 600 feet on a steep mountain near the Leet-head. In the narrow Barranca at the foot of this place the Mescal plant is extensively cultivated. Bananas and oranges abound, and the shingle roof of a small hut, in which the Vino Mescal is distilled, was just visible amongst the beautiful wilderness of the place.

CHAPTER VI.

Mines of Bolaños—Woods—Leet or Canal—Former expense of draining the Barranco Mine—Town of Bolaños—Animals, Fruits and Vegetables—Native Indians—Guicholas.

The mines of Bolaños, now the object of such important speculations, and whence such valuable fruits may rationally and confidently be anticipated, are situated along the eastern level of a deep Barranca, at the foot of precipitous cliffs which hem it in on that quarter. The shafts and levels rarely occur at a perpendicular height of above fifty fathoms up the steep mountain's side, and the buildings of the *Grass-works* lie almost hidden amongst close thickets and delightful scenery. The veins for the greater part are in porphyry, intersected by strata of crumbling steatite, or soapstone, of a variety of colours—red, gray, black, green, brown, yellow, mottled, and a very pure white. It is owing to this "soft ground" that many of the old workings have suffered considerably. Removed from the mineral veins are beds of a fine white freestone, admirably fitted for architectural purposes; and the towering " Aguila" and " Bufa," with other cliffs, which in some instances rise 1500 feet above the wooded slopes of 1000 more, are also of the same material, having for the most part their strata dipping 45° to the north. Limestone is found at no great distance to the south, and the soil in all parts is rich and abundantly productive, but ill-cultivated, and hitherto poorly watered. I did not observe that in any instance the river had been employed for the purpose of irrigation.

At about seven miles to the westward of Bolaños is a ridge of rugged mountains, clothed to their summit with fine firs. Their height above the river was ascertained by Captain Vetch to be 5000 feet; and I conceive that they are a part of the range which the Baron de Humboldt designates as the " Sierra Madre." Rich as they are in large timber, it is rendered almost useless, as the mountain roads are so precipitous and bad

that it is almost impossible to transport spars of
any magnitude along them. The present supplies
of wood requisite for heavy machinery are therefore procured from a distance of ten or twelve
leagues; but for ordinary purposes, the wood
around the mines would be inexhaustible, with
a hundred times the probable demand. The
immediate valley and steep slope from the foot of
the eastern cliffs runs north and south for about
four miles, without a turn of any importance; and
the general course of the metallic veins is N.N.E.
and S.S.W., slightly deviating occasionally to the
east or westward.

The whole of the mines in the Barranca of Bolaños are for a series of years the property of the
English Company of that name, although the
owners and lessors are various. The line to which
the mineral riches are confined is divided for
greater facility of accounts into three portions:

1. Northern Bando.

2. Entermedio.

3. Southern Bando.

The Northern Bando consists of three distinct *sets*, which are named Concepcion, Tepec, and Camichin: of these the mine of Concepcion is the most northerly, having its principal shaft, which is a considerable way up the woody hill, cleared to the depth of forty-six varas *. I entered this mine by a good socabon, (driven through a very compact porphyry,) which, with a cross cut on the lode, had been since the period of the former workings entirely choked with attle, but has now been completely cleared; and hands are driving on an old working, to cut the lode, which, with a dip to the N.W., runs N. 60° E., and is from three to six feet in size. At this period I saw but little promise in the ores.

The temperature of the mine was 93°, while the outer air was 83°. Tepec contains the old and new shafts of Santa Fé, with those of Santa Cruz and San Cayetano.

I visited the workings of *New* Santa Fé, those of

---

* At the depth of thirty varas it has been floored, and a cross cut has been commenced, south, to the lode.

the *Old* being higher up the mountain, and quite choked, descending by a shaft of thirty fathoms, very soundly timbered, and entered a new cut of 122 feet, which has reached the lode, above the back of which an upward rising has been made at an angle of 45°, to try for the old workings. This cut is through a soft soap-stone stratum, while the lode beneath is hard and difficult, and rendered more so till the communication with the old workings is completed, by the extreme closeness and want of air in the mine. There can be little doubt that this upward cut will soon reach the old workings, when there is every reason to expect good fruits; since, were ventilation once obtained, the base of the cutting is the lode itself, and can be worked at pleasure.

From the extreme closeness of the air (96°) an extra sum is paid to the Barrateros. It is fortunate that the nature of the present cuttings is such as not to require the process of blasting,—an operation which could scarcely be performed in so confined and exhausting an atmosphere.

The next mine on the set is Santa Cruz, which has not been deemed worthy of trial. To this succeeds San Cayetano, to which I descended by an old shaft which has lately been cleared and timbered to the depth of twenty fathoms. The workings in this mine were formerly very extensive; and the miners have recently arrived at large arches of old *labores*, very rich in lead, with good indications of silver, among which I found some of that kind called Petanque. The old works have been very irregularly driven, so that much of the lode, which varies from eight feet to eighteen inches in width, has been left untouched. The ground is very hard, but there is every reason to expect good returns when proper trials are made for discovery. The chief part of the lead is large-grained, and of that kind called *potter's lead*, which is unproductive of silver; but there is also ore of a finer grain, which promises well. Abundance of each kind may be procured.

This mine bears an excellent character; and as it is very shallow, I should conceive that much virgin ground may be advantageously explored, since the mine, in common with all in Tepec, is quite dry, and no adit has been either cut or required.

Camichin consists merely of a mine and socabon of the same name. Much of the old workings have been cleared out; but the works are stopped for the present.

The socabon is excellent, driven through porphyry; but all the workings from it appear likely to be unproductive, since the extreme hardness of the rock would render trials for discoveries difficult and expensive. Here the temperature was only 76°, while the outer air was 89°.

The Intermedio contains the Carolina, Socabon de los Negros, and the shafts of San Juan (new), and America.

I first entered by the Socabon de los Negros, the mouth of which being ill-contrived has been remedied by a small under cut, with a clack-door, as the river has sometimes risen in the rainy season so as to enter the adit. The present entrance is therefore to be stopped, and communication is to be effected higher up the hill. A cross cut of eighteen fathoms has been completed to San Juan's, or Taylor's new shaft, which has been sunk to the depth of forty varas, and is the only one over which a Cornish whim has been erected. Men were at this time working here on a new level, running S. 9° W., in very hard ground, containing a narrow lode of lead and silver, neither of which were very productive. Another set were also driving, N. 20° E., in hard ground, and had not reached the lode. In the first working the temperature was 90°; in the second, 92°. The former *labor* has been extensive, and the lode has in many places been entirely cut away.

Carolina has been in work, but is at present stopped. Four fathoms of new ground have been opened, and abundance of potter's lead, two feet wide, can be easily attained. The temperature was here 94°, although no workmen were under-ground.

America shaft, which was entirely choked, has been cleared down to water; but as yet, nothing of interest has been discovered.

The whole of the mines of the Northern Bando are dry, shallow, and comparatively but little worked. Santa Fé and San Cayetano promise to become productive when active workings shall be commenced upon them. The other mines in certain spots also look well. The ruin of these places has been very great; but I could not have believed it possible that so much of the old workings should so speedily have been cleared. Very extensive timbering has been accomplished, and new levels have been driven,—to the great credit of the officers in direction of the *Negociacion*.

The Southern Bando has two *sets*, Barranco and Laureles. I entered only the former celebrated mine, by its magnificent socabon, which runs a considerable way in before it branches off to the Guadalupe shaft and to that of San Diego; both of which are clear to the adit level, but choked beneath.

There is little to be seen beyond this, owing to the ruined state of the mine, which at this part, being above the level of the river, is perfectly dry.

Traces of the rich vein were distinctly visible. Above-ground, some distance up the steep brow of the hill, are the immense and magnificent remains of the great Hacienda, built of freestone, and having the inclosures and pillars for the numerous malacates which were once employed, still standing.

Their labours were all devoted to draining the great Guadalupe shaft, which is still open to the depth of thirty fathoms, and has now no timber remaining. San Diego has been a very fine shaft. Zapopa is filled up; but that of San Vicente is good at the mouth, and open to some depth.

The Cocina shaft, where it is intended to place a water-wheel, has been cleared to eight fathoms, and is situated on the low ground between the Hacienda del Barranco and the town of Bolaños, which lies to the northward.

I did not visit Laureles, which lies south of the Barranco mine, and has been worked to the depth of forty varas. It has the shaft of Renovales, which is now so choked up that little or no communication can be had with the former workings. The river, which runs hence to the southward and joins the Rio Grande or San Cristobal, was fordable at the time of my visit, rushing over a somewhat rugged and stony bed; but its present appearance will be materially altered when its waters are turned to the great object to which all the extensive works are now so rapidly advancing. It is the plan of the Company to bring the stream by a capacious Tarjéa or leet* to a large water-wheel which will be erected at the shaft of Cocina, whence flat rods will be carried up a small hill, and applied to the purpose of pumping the water from the celebrated mine of Barranco by the Guadalupe shaft. To accomplish this important object, 6740 varas of ground of various descriptions was to be opened. In some instances, drivage through very hard rock has been effected; and in others, either through freestone or in open cuttings. At the time of my visit, half of this great work had been accomplished; and in twelve months, if suffi-

* Canal.

cient outlay be afforded, the whole leet may be completed to its intended termination. The economy now so necessarily adopted, in consequence of the recent distresses in England, has materially checked the progress of the operations at Bolaños, where a boldness of outlay is of all places the most to be recommended.

At the period of my visit, much anxiety was felt as to the building up and securing the soft open cuttings of the leet, no clay proper for bricks having been discovered; but very shortly subsequent to my departure this very necessary material was found, and abundance of very good bricks have been made. The walls can now speedily be carried up, while the underground drivage is at the same time advancing.

The ores, which in small and rather unproductive quantities had been taken from the various workings, were deposited in a large walled yard attached to the Hacienda of Tepec. Shops and stores for the mechanics, with substantial residences and offices for the mining captains, were nearly completed, on a central eminence; and the whole establishment had already assumed an appearance creditable to English skill and activity.

The chief attention and labour of the agents of the Company continue to be directed to the grand and important task of bringing on the leet: the whole expense of which will certainly fall short of two years' expenditure in the former unsuccessful attempts to drain the great Barranco mine by malacates,—an operation which was persevered in with such ardour, as to give the strongest proof of the riches which are known to exist in the mine.

The following is a translation of a very curious list of the expenses incurred by the last adventurers*, and will show the great importance of the Barranco mine.

* From an old paper addressed to the mine-holders in 1795, submitting to them at the same time a plan of economy which was not then followed. The document is valuable, as showing the exact amount of the drainage alone, without the salaries of superior officers, or expenses incidental to the mines themselves.

" Expenses of the Desague (drainage), on the method which has hitherto been observed.

" To maintain the mine that the water does not rise
to above two Cañones (levels) above the Planes
(bottoms) of the Barranco, forty-four Malacates
(horse whims) have been considered requisite;
for which the following expenses were annually
required:

" For this purpose, 2,200 mules, fifty to
each malacate, have been necessary.
For their subsistence one year with another, with alterations and diminutions
in the prices of maize and forage at two
reals *per diem* for each animal, the annual expense is . . . .

	Dollars.
" For this purpose [...] nual expense is	200,750
" 176 Drivers by day and night, at one dollar each	64,240
" 88 Assistants by ditto, at five reals each	20,075
" 10 Watchmen in the five ' tiros de desague' (shafts at which the water is drawn), at eight dollars a week each	4,160

" 88 Boteros (fillers of the water-bags), *Dollars.*
night and day, at one dollar each   .   32,120

" 6 Señeros (those who make signals to
hoist lower, &c.), at one dollar each   2,190

" 16 Desaguaderos (who receive and empty
the water-skins), on a weekly salary
amounting to   8,320

" It is ascertained that each Calabrote
(whim rope) lasts one month, one with
another; these are each composed of
forty-five Sogas (small lines), which cost
eight reales and a half the dozen. The
total cost is   25,202

" To repair the whim ropes, &c., mend and
make reins for mules, and for casual-
ties which occur to each malacate, fifty
dozen Sogas are required weekly in the
five Tiros (shafts)   2,762

" For water-bags (Botas), and mending
those which break, 120 Vaquetas (cow-
hides) are required *per mensem ;* these,

one with the other, cost four dollars. In
a year . . . .   *Dollars.*
   5,760

" To the above are to be added the expenses
of houses, machinery, soap to ease the
friction of the working pieces, traces,
whips, water-skins, &c., malacates, repair of sheds over them, mules which
die, salary of the mule-keepers, Ocote
(candle-wood) to light the works, with
many smaller expenses which cannot be
enumerated; for this, fourteen dollars is
allowed for each malacate weekly. Making a total annual amount of . .   32,032

Total expense of drainage (Desague) . 397,612"

The town of Bolaños, which according to Captain Vetch is at an elevation of 3101 feet above
the sea, now contains about a thousand souls,
having nearly doubled its population since the
Habilitacion of the mines by the English; but in
its days of greatness and splendour the population
is said to have exceeded thirty-five thousand. Instead of being a cluster of miserable mud huts, as

people have imagined in England, it has on the contrary been a beautiful little town; and, were it not for the extreme sultriness of the weather, would be a most delightful residence. Agues and fevers are very prevalent, even amongst the natives; and their consequences are long felt by invalids, who recover their strength but slowly.

Amidst the hot airless days of the dry season, there is however a delicious relief constantly at hand, in the then clear river; in which indeed throughout the year the people of both sexes bathe night and day, with very few scruples as to publicity. The water was turbid at the period of my visit, the rainy season having recently concluded; but sweet and good, producing fish in tolerable plenty, although in no great variety. I only heard of the Bagre, Trout, Boquinete, the last of which is said to be bony and insipid; the Sardina, a small fish; the Mojarra, a flat species; minnows, and a large sort of crayfish.

Alligators have also occasionally made their appearance even at this remote place; and one at no very distant period was caught alive, measuring four varas (about eleven feet), near the Hacienda of Tepec.

The woods on the borders of the river are said, although I confess to having seen no proofs of it, to be the resort of wild turkeys, deer, hares, rabbits, and other varieties of game: but I know that the common green parrots, with parroquets, and a very large variety somewhat resembling the Indian maccaw, and named Guacamalla or Papagallo, are also abundant at no great distance.

Snakes, scorpions, and other venomous creatures abound in the houses as well as the woods; and the variety of iguanas and lizards amongst the rocks and thickets is very remarkable.

The articles of subsistence are somewhat dearer here than on the Table Land, evidently in consequence of the difficulties of transport. Vegetables however are sufficiently plentiful, but only of the most common species, although better kinds and greater varieties would undoubtedly thrive well. Fruits, which are also cheap and tolerably good, are grown in various parts of the neighbouring Barranca. These chiefly consist of pines, peaches, apricots, apples, pears, aguacates, plantains, bananas, guayavas, grapes, quinces, musk and water melons. The delicious Chirimoya, two species of the Garambuyo plum, one yellow the other red; Ciruelo, or the service; Garlāmo, a black wild berry; and the sugar-cane, which being eaten raw in great quantities may also be called a fruit. To all these, however, must be added the luscious " Pitaya," the fruit of a tall tree of the Tuna family *, which is peculiarly fine at Bolaños. The vegetables are; pumpkins of various kinds, onions, beans, pease, garvanzas, Lenteja, which resembles the European lentil, Frijoles (long French beans of various colours), cabbage, chile of several varieties, ochra, camote, chayote, and probably some inferior wild vegetables, of which I did not hear the name. Maize is cultivated in small quantities, and the maguey is seen here and there; but not so abundantly as the Mescal, which is consumed in the fabric of a fiery spirit called Vino Mescal.

* A species of Cactus, called also Organo from its figure.

Bolaños appears to have been the original Indian name of the Barranca, and portions of several tribes still live in its immediate vicinity; but of late years the distinction of nations has given way to a more general admixture of tribes, and it is no longer possible to distinguish the *Indios Mensos** of this part of the country by their former national divisions. The Guicholes are in fact the only neighbouring people who still live entirely distinct from those around them, cherishing their own language, and studiously resisting all endeavours to draw them over to the customs of their conquerors.

These and other Indians make the shores of the Pacific about six days' journey from Bolaños, at their travelling rate of seven or eight leagues per day; but the route is difficult and mountainous, and never pursued by any but themselves: the

---

* "Tame Indians," so called in contradistinction to the *Indios Bravos* or "Wild Indians" who have not embraced the Christian religion, and are generally at enmity with the whites.

VOL. I.

road to San Blas by way of Guadalaxara, being preferred by Europeans or Creoles.

The Guicholes are settled in the village of San Sebastian, which lies eighteen leagues to the westward of Bolaños, and two days and a half from the spot whence the salt is procured, which is called Quaristemba. They live in small scattered communities, but have also two villages, one named Santa Catalina, twelve leagues beyond San Sebastian, and the other called San Andres Coasmatl. The whole of the country however, between Bolaños and the Pacific, is very little known.

Near a village named San Martin, situated a long day's journey in the mountains to the southward, there is said to be a cave containing several figures or idols in stone; and had I been master of my time, I should most assuredly have visited a place which is still spoken of with much interest by the natives. All the antiquities I was enabled to procure at Bolaños, by offering rewards, were three very good stone wedges or axes of basalt; and on its being known that I would purchase curiosities, a man came to inform me that at a long day's journey could be found *Huesos de los Geutiles*, or Bones of the Gentiles, of which he promised to bring me some if provided with mules, since their size was very great: he must however have been absent three days on this errand, and as my stay was limited, I could not give him the commission, although anxious to obtain what I had reason to believe were the bones of elephants or mastodons.

END OF THE FIRST VOLUME.